Obesity and its Impact on Health

Methil Kannan Kutty • Asita Elengoe
Editors

Obesity and its Impact on Health

Springer

Editors
Methil Kannan Kutty
Department of Pathology
Faculty of Medicine, Lincoln University College
Petaling Jaya, Selangor, Malaysia

Asita Elengoe
Biotechnology Department, Faculty of Science
Lincoln University College
Petaling Jaya, Selangor, Malaysia

ISBN 978-981-33-6407-3 ISBN 978-981-33-6408-0 (eBook)
https://doi.org/10.1007/978-981-33-6408-0

This Springer imprint is published by the registered company Springer Nature Singapore Pte Ltd.
The registered company address is: 152 Beach Road, #21-01/04 Gateway East, Singapore 189721, Singapore

Foreword

The revolution in health in the last century has led to a paradigm shift in disease morbidity. The consequent demographic transition has led to an epidemiological transition from communicable diseases to chronic diseases. Countries are facing insidiously growing phenomena where nations are grappling with the double burden of disease which is a coexistence of communicable and chronic diseases. Obesity is the major risk factor for a variety of chronic diseases such as diabetes mellitus (DM), cardiovascular diseases (CVD), hypertension, endocrine disorders, endometritis, and cancer.

Over the last 40 years, the global prevalence of overweight and obesity rates has doubled. Approximately, a third of the world's population is now classified as overweight or obese. Obesity rates have increased across all ages. It ravages both the developed and developing nations. Obesity could be viewed as the merged impact of lifestyles, environmental factors, and individual genetic predisposition. CVD is the major cause of death in the USA and most other countries. Obesity is complex and multifactorial but largely a preventable disease. Hence, it should be dealt with more aggressively to reduce the risks of this problem.

The most common estimate of obesity is the physical measurement by body mass index (BMI) that relates the height and weight of an individual. BMI is defined as a ratio of weight (kilograms) and height (square meters). A BMI between 20 and 25 kg/m^2 is normal and associated with the lowest mortality, whereas a BMI of 25–30 kg/m^2 is considered overweight. World Health Organization declared that changes to lifestyle and an increase in the availability of energy-rich foods are important contributors to global obesity. It is also becoming prevalent in Asia countries, requiring urgent action to combat this epidemic.

This book on "Obesity and Its Impact Health" by Professor Dr. Methil Kannan Kutty and Dr. Asita Elengoe is a timely reviewed effort to fill this gap. It includes 12 chapters. Chapter 1 describes the introduction of obesity and the chemical components which enhance fat deposition in the human body. Chapter 2 deals with the economics of obesity. Chapter 3 discusses the impact of obesity on the skin. Chapters 4, 5, and 6 describe how obesity causes endometriosis, hypogonadism, and the obesity effects on the cardiovascular system. Chapter 7 focuses on the impact of obesity on vaccination. Chapter 8 features obesity-linked diseases (comorbidities). Chapter 9 emphasizes imaging as a tool for measuring body fat. Chapter 10 focuses on extrapolation from clinical trial to practice: current pharmacotherapy on obesity. Chapter 11 deals with the nanotechnology approach in the context of obesity. Chapter 12 describes the computational approach in drug development for obesity. This book will prove to be highly beneficial to the academicians, researchers, and students.

Melaka Manipal Medical College,
Faculty of Medicine
Melaka, Malaysia

Jayakumar Gurusamy

Preface

Obesity is an escalating global health problem of colossal magnitude and its incidence is mounting stunningly in several nations including industrialized countries. It has impacted the health of all persons regardless of age, gender, race–ethnicity, and socioeconomic status besides many other facets. Obesity is determined by the proportion of body adipose tissue. It is a medical problem that increases the risk of other diseases and health problems such as heart disease, hypertension, diabetes mellitus (DM) type 2, certain cancers, male and female reproduction system-related diseases (e.g., endometritis, polycystic ovarian syndrome (PCOS), etc.), kidney-related disorders, osteoarthritis, and digestive problems. A body mass index (BMI) over 25 is considered overweight, and over 30 is obese. The issue has grown to epidemic proportions, with over four million people dying each year as a result of being overweight or obese in 2017 according to the global burden of disease. Rates of overweight and obesity continue to grow in adults and children.

There are many factors that cause obesity such as an unhealthy diet (e.g., fast food like KFC, McDonald, Pizza Hut, etc.), liquid calories (e.g., alcohol and soft drinks), lack of physical activity, family inheritance, certain diseases such as Prader–Willi syndrome and Cushing syndrome, social and economic issues, age, stress, lack of sleep, and pregnancy. Currently, many studies are in progress to discover novel mechanisms in the control of the initiation and evolution of the pathological state of obesity.

This book presents 12 chapters. Three of the contributors wrote on obesity to current perspectives on pharmacotherapy, nanotechnology, and computational biology tool. These novel approaches can be a feasible strategy that can combat obesity and prevent its comorbidities. There is a chapter on obesity-linked diseases (comorbidities). We have also attempted to revisit the obesity and skin, obesity and endometriosis, hypogonadism, obesity and cardiovascular system, and impact of obesity on vaccination and imaging as a tool for measuring body fat. The understanding of the molecular mechanisms regarding the development of obesity is vital to design anti-obesity drugs. Economics of obesity seamlessly interlocks to enhance the main theme "obesity compliant with the modern era and concept of borderless knowledge."

We highly appreciate our contributors for accepting our invitation and shared their knowledge and research. We produced an excellent quality book with a good teamwork effort. We would like to thank the Springer Nature team, Bhavik Sawhney and John Martyn for their kind cooperation at every stage of the book production.

Petaling Jaya, Selangor, Malaysia Methil Kannan Kutty
Petaling Jaya, Selangor, Malaysia Asita Elengoe

Acknowledgments

> The important thing to recognize is that it takes a team and the team ought to get credit for the wins and the losses. Successes have many fathers, failures have none.—Philip Caldwell

First and foremost, I (Prof. Dr. Methil Kannan Kutty) would like to thank wholeheartedly my wife Innira for all her work that has kept my profession and our family going all these years. This book like all my other publications is a testament to her constant sacrifices.

This book acknowledges wholeheartedly and gratefully all my colleagues for their valuable contribution. I cannot resist my temptation to applaud the enthusiasm of these experts that catalyzes my academic passion and catapults me beyond my limitations.

My special thanks to Dr. Asita Elengoe for all her unstinted cooperation at all phases of the preparation of the book.

I (Dr. Asita Elengoe) would like to express my sincere gratitude to GOD for His blessings to complete this book successfully. I sincerely thank Prof. Dr. Methil Kannan Kutty for his initiative thought to start writing up this book. My sincere gratitude also goes to my dear friends who supported me during the completion of this book. Last but not least, I would like to thank my mother, Thavamani who supported me spiritually throughout my life.

About the Book

Obesity has not only become a pan-global epidemic but also it is the fifth leading risk factor for global mortality and it appears to be an escalating problem, as menacing as cancer. Hence, we are prompted to review and relook at the current trends in the literature on this subject. The contributors have provided the latest information on obesity with particular reference to the current perspectives on pharmacotherapy, nanotechnology, and computational biology. Besides, there are chapters on impact of obesity on endometriosis, hypogonadism, and obesity-linked diseases (comorbidities). We have also attempted to revisit the obesity and skin, obesity and cardiovascular system, and impact of obesity on vaccination and imaging as a tool for measuring body fat. The Economics of Obesity: A Public Policy Perspective is also included in this book. As we had contributed to a book on Obesity and Cancer, this aspect has been deliberately omitted in the present book. Although facets appear to be cloistered in different chapters, we have endeavored to thread together to provide a comprehensive review that will be meaningful in this era of borderless knowledge.

Contents

About the Editors

Methil Kannan Kutty is working as Professor Pathology in Faculty of Medicine, Lincoln University College, Malaysia. His qualifications include MBBS (Madras), MD (Lucknow), FRCPath (London), and FRCPA (Australia). He has served as Professor of Pathology in Universiti Islam Malaysia, Universiti Kebangsaan Malaysia, UITM, International Medical University, Royal College of Medicine and King Faisal University, Dammam, Kingdom of Saudi Arabia. He was appointed as visiting professor at Madras Medical College for 5 years. He was the recipient of Sir David Galloway memorial lecture award from the Academy of Medicine Singapore University in 1974. He has to his credit more than 160 publications in peer-reviewed journals. He is the author of seven books on various subjects such as (1) Menstrual Disorders Menarche to Menopause, (2) Endometriosis: Past, Present and Future, (3) Endometriosis Current Concepts and Future Directions, (4) Obesity and Cancer Current Concepts, (5) Dysfunctional Uterine Bleeding Current Concepts, (6) Obesity Current Trends, and (7) Insight in Pathology. His research interest is in: (1) rhinosporidiosis, (2) sarcosporidiosis, (3) endometriosis, and (4) obesity.

Asita Elengoe is currently the head of Biotechnology Department at Faculty of Science, Lincoln University College, Malaysia. She obtained her PhD in Bioscience (2015) and Master of Science in Biotechnology (2010) from Universiti Teknologi Malaysia. She received her Bachelor of Science in Biotechnology (Hons) (2008) from Universiti Tunku Abdul Rahman. Her research interest includes cancer, gene therapy, computational biology, natural products, and medicinal plants. She is a member of Royal Society of Biology (MRSB), UK. She is the editor for SCIREA Journal of Biology and assistant executive editor for International Journal of Advancement in Life Sciences Research (IJALSR). She is the reviewer for indexed and non-indexed journals. She is actively involved in research and publications. She has published articles in indexed and non-indexed journals. Three book chapters were already published in "Software and Techniques for Bio-molecular Modelling," "Anticancer Plants: Mechanisms and Molecular Interactions," and "Plant Derived Bioactives" books. She received an award for the "Young Researcher in Biotechnology" at International Healthcare Awards 2018, Chennai, India. Currently, she carries out research on drug designing using computational biology approach.

Introduction 1

Prameela Kannan Kutty and Methil Kannan Kutty

Abstract

The global obesity rates are escalating alarmingly. It is determined by the proportion of body adipose tissue. Its colossal magnitude is a pan-global problem seriously impacting all nations irrespective of geographic, demographic, socio-economic and other factors. The grade of obesity is determined by body mass index (BMI). This is classified by World Health Organization (WHO). A person who has BMI of more than 30 kg/m^2, he or she is considered obese. It seems that obesity merits to be rechristened "adiposopathic-multifanged malady" as it unquestionably connotes a combination of pathogenesis as well as comorbidities such as diabetes mellitus, hypertension, dyslipedemia, hyperlipedemia, respiratory problems, kidney-related diseases, cardiovascular pulmonary disturbances, osteoarthritis, sleep apnoea and cancer. The multifacetted profile of obesity and its related conditions make it imperative to revisit this serious health problem which extends beyond what the eyes can see. In this book chapter, we will discuss the myokines, adipokines, BDNF, leptin, vistafin, resistin, ghrelin, chemerin and apelin roles in obesity.

Keywords

Obesity · Myokines & adipokines · BDNF · Leptin · Vistafin · Resistin

P. K. Kutty
National Defence University of Malaysia, Kuala Lumpur, Malaysia

M. K. Kutty (✉)
Department of Pathology, Faculty of Medicine, Lincoln University College, Petaling Jaya, Selangor, Malaysia

M. K. Kutty, A. Elengoe (eds.), *Obesity and its Impact on Health*,
https://doi.org/10.1007/978-981-33-6408-0_1

1.1 Introduction

Obesity is an escalating global health problem of colossal magnitude and its incidence is mounting stunningly in several nations including industrialized countries. It is by no means an exaggeration to define obesity as a global scourge that has impacted the health of all persons regardless of demographics and other factors. According to some, obesity is determined by the proportion of body adipose tissue. For men, this percentage is one that exceeds 25% and it is 33% for women. Obesity is perceived as a disturbance of metabolism associated with a background state of low grade inflammation (Monteiro and Azevedo 2010; Donath and Shoelson 2011). This concept fails to enlighten the whole clinical profile of obesity; currently, there are efforts to more clearly understand the pathogenesis of the initiation and evolution of obesity.

It is pertinent and is of consequence to allude to the historical background of the definition of obesity. Strangely in the early era, the term "obese" was deemed as uncomplimentary and was seldom used (Nuttall 2015). Data on weight for height and tables for individuals who were underweight and overweight were reviewed at different time periods. As an alternative, Keys et al. (1972) recognized the importance of data comparing weight for height; and promoted the Quetelet Index in population-based studies. They referred to it as the body mass index (BMI). Thus, Quetelet Index is the body weight in kilograms divided by the square of the height in meters. However, Keys et al. (1972) indicated even the BMI somewhat poorly signifies an individual's percent of body fat. The WHO classification is widely used (World Health Organization (WHO) 1995).

Albeit diverse categorisations of grades of obesity, the most broadly recognized classification is the BMI determined by the WHO. A BMI of 25–29.9 kg/m^2 constitutes overweight, while obesity is defined by a BMI of 30–39.9 kg/m^2 and morbid obesity by a BMI ≥ 40 kg/m^2.

It seems that obesity merits to be rechristened "adiposopathic-multifanged malady" as it unquestionably connotes a combination of pathogenesis as well as a spectrum of comorbid conditions which include hypertension, diabetes mellitus, dyslipedemia, hyperlipedemia, cardiovascular pulmonary disturbances, sleep apnoea and cancer (Zhiyong et al. 2016; Michaela et al. 2016). In addition, obesity is linked with the risk of gall bladder disease, hyperuricemia, gout and osteoarthritis (Khaodhiar et al. 1999). Obese individuals, especially those with abdominal adiposity, are at escalated risk for coronary artery disease (CAD). In 1998, the American Heart Association included obesity in the list of major risk factors for CAD (Eckel and Krauss 1998). It is noteworthy to highlight that certain cancers are also linked with obesity, comprising colorectal (Alireza et al. 2007, Kutty et al. 2013) and prostate cancer (Allott et al. 2013) in men and endometrial (Michaela et al. 2016), breast and gall bladder cancer in women. Overweight and obesity are also linked with an escalated hazard of solid tumors and hematological malignancies (Renchan et al. 2008). These have been dealt with in detail in the book entitled "Obesity and cancer". Obesity has a link with a gamut of hepatic disease known as non-alcoholic fatty liver disease (NAFLD) or non-alcoholic steatohepatitis (NASH) (Schwenger et al. 2019). The prevalence of NAFLD has escalated and continues to increase. NAFLD is associated with simple steatosis but may also be linked to more severe forms of the disease which can progress to cirrhosis and malignancy. NAFLD is regarded as the hepatic expression of the metabolic syndrome and its complications. The pathophysiology of NAFLD is linked to adipose tissue alterations with immunity and intestinal barriers acting as initiators and prom and escalated risks of a spectrum of widespread malignancies (Dietrich and Hellerbrand 2014). An important facet that needs emphasis is that with increasing obesity in children and adolescents. They are prone to develop risk factors for chronic diseases frequently seen much later in life, such as dyslipidemia, hypertension, and hyperinsulinemia (Freedman et al. 2007). This is illustrated by the increasing population of obese children and adolescents who are now being diagnosed with type 2 diabetes disease that was practically absent globally a few generations ago (Fagot-Campagna et al. 2000). Likewise, there is proof in support of obesity in children and adolescents prone to exhibit the advance of cardiovascular disease (Choukem et al. 2020).

It is also useful to consider the link between obesity and the prevalence of leukemia and non-Hodgkin's lymphoma especially diffuse large B cell lymphoma (Marshall and Lichtman 2010). A study revealed an increased BMI is linked with higher relative risk RR of diffuse large B-cell lymphoma (DLBCL) regardless of sex with a linear association between BMI and DLBCL incidence (Castillo et al. 2014). An elevated BMI is correlated with a higher risk for myeloma, but despite the correlation between obesity and malignancy the causal link is inconclusive. The impact of obesity can alter cell mutation rates; and interfere with gene function and DNA repair, prompt epigenetic changes, supporting the induction. Obesity could engender a micromelieu which could trigger the emergence of dormant pre-existing clones (Dorantes et al. 2020).

There are exogenous causes of obesity such as drugs namely corticosteroids (Emily and Dunford 2016). High cortisol levels increase appetite and derange lipid and glucose metabolism resulting in a clinical picture that depicts mainly truncal distribution of fat deposition (Fardet and Feve 2014). Prolonged usage of oral corticosteroids leads to exogenous Cushing's syndrome. Increased incidence of obesity (Ng et al. 2014) and increased corticosteroid usage (van Staa et al. 2003) are indeed important concerns and causes of general weight gain today.

Obesity may also be linked to hypothyroidism being secondary to thyroid dysfunction. It may be a two-way process whereby alterations in thyroid-stimulating hormone (TSH) could also possibly be explained by changes associated with obesity. Links between obesity, autoimmunity, with actions related to the adipocyte hormone, leptin, have been suggested by Debmalya and Moutusi (2016).

Thyroid hormones are crucial determinants of metabolic processes such as basal metabolic rates; thermogenesis; sugar and fat metabolism as well as food consumption and the beta-oxidation of fats. There is the support that subclinical hypothyroidism is linked to body weight changes (Danforth et al. 1979).

Ionic channel disruptions may add to the pathogenesis in obesity. Ionic channels involving sodium, potassium and calcium chloride coupled acetylcholine receptors and receptor proteins may contribute to ionic disturbances observed in obesity (Vasconcelos et al. 2016).

It is of interest to review the basis of the involvement of potassium channels linked to early taste perception (Vasconcelos et al. 2016), oxidative stress and inflammatory cytokines in the pathogenesis of obesity and its comorbidities (Ramírez et al. 2016; Prasenjit and Sushil 2015; Mario et al. 2018).

Transient Receptor Potential Canonical 5 (TRPC5) represents a Ca^{2+} permeable non-selective cationic channel which adversely controls adiponectin sans any effect on leptin in mice fed chow diet (Beech 2013). Adiponectin is a major anti-inflammatory mediator. Therefore, it was assumed to have an influence of TRPC5 on the inflammatory condition of atherosclerosis. A study of Atherosclerosis in the aorta of $ApoE^{-/-}$ mice fed western-style diet revealed surprisingly DNT5 instigated a very substantial decrease in body weight gain and condensed adipocyte size following 6 and 12 weeks of the western-style diet. There was hardly any effect on steatosis and circulating lipids. The data propose that TRPC5 ion channel permeation which has little influence on atherosclerosis or steatosis surprisingly a major effect on weight gain (Baptiste et al. 2019).

Increased fat accumulation in humans has also been correlated with oxidative stress biomarkers (Keaney et al. 2003). Reactive oxygen species (ROS)-induced changes in fat and proteins are clinical biomarkers in humans (Fabio and Ilaria 2017).

Rodent experiments indicate that adipose tissue produces reactive oxygen species which enter circulation and these substances possibly negatively impact a number of organ systems (Furukawa et al. 2004).

Therefore, the pathogenesis of obesity is indisputably multifactorial with the chief causal factors being unsuitable diet and lifestyle while genetic factors have been shown to modulate the phenotypic expression of obesity (Jana et al. 2012).

1.2 Myokines and Adipokines in Obesity

It will be consequential and enlightening to describe briefly the relationship between myokines and adipokines in obesity. In this regard, myokines refer to muscular exercise factors that are generated by skeletal muscles as a reaction to exercise that is released into the blood. These factors include ANGPTL4, CCL2, CX3CL1, IL-6, IL-8, IL-15, myostatin and SPARC among others (Rosanna 2019). Human obesity is linked with amplified myostatin expression and plasma myostatin levels. The secretion of myostatin from myotubes in muscle biopsies is augmented in obese compared with lean women (Allen et al. 2011). These may be treatment targets in obesity.

White adipose tissue is a dynamic endocrine organ. It secretes signalling proteins called adipokines which include cytokines, complement linked proteins, vasoactive peptides and biological mediators. Adipokines have important autocrine or paracrine functions and contribute to the much obesity-linked comorbid conditions.

Metabolic alterations linked with weight gain, especially visceral obesity, coupled with an inactive lifestyle could lead to the malfunction of adipose and muscle tissues. This would trigger a series of changes encompassing not only insulin resistance and low-grade chronic inflammation but also the abnormal release of adipokines and myokines. The regular interplay between adipokines and myokines reveals key links in developments such as the regulation of lipolysis and insulin sensitivity (Maria 2013).

Knowledge of the mechanisms of the adipokines and myokines' interaction with cancer physiopathology will enhance the salutary approaches against cancer cachexia.

Myokines and adipokines are synchronised by inflammation- linked proteins as well as autocrine and paracrine interconnections in muscle and fat. This may function in two usually opposite directions; for instance, there is a role of IL-6 in skeletal muscle dissipation and its restoration (Belizário et al. 2016).

Before we discuss obesity per se, it is essential to have a better understanding of the metabolic disturbances and the grade of related cardiovascular and metabolic risks. Obese individuals manifest differences in their body fat distribution. It is well known that abdominal obesity is beset with higher susceptibility to develop diabetes and future cardiovascular problems than peripheral obesity. The dissimilarities between adipose tissue confined to subcutaneous areas (SCAT) and visceral adipose tissue (VAT) is existent in the abdominal cavity. Relative to the profile of SCAT some features of VAT are as follows. It's predominant in the mesentery and omentum and thus it is drained straight into the liver via the portal system. It is very cellular and vascular with significant inflammatory and immune cells. It has an enhanced population of large adipocytes due to lesser pre-adipocyte differentiating potential. It has also increased the number of glucocorticoid and androgen receptors. The association with adipocytes has higher metabolic activity; and this increased the sensitivity to lipolysis, as well as greater insulin-resistance than SCAT adipocytes. It has a greater propensity to generate free fatty acids and to absorb more glucose. VAT exhibits a higher sensitivity to adrenergic stimulation. It is vulnerable to a higher mortality rate (Vasconcelos et al. 2016).

1.3 BDNF

A brief allusion to Brain-Derived Neurotrophic Factor (BDNF) and its significance in obesity is pertinent in this context. Factors implicated in the equilibrium of energy disbursement regulation of lipid and glucose levels; and cardiovascular homeostasis, symbolize a novel family of metabotropic factors (Fiore et al. 2014) such as BDNF (Chaldakov 2011).

Many cells including neurons, adipocytes, immune cells and others synthesise BDNF (Nakahashi et al. 2000). The multidimensional function of BDNF encompasses neurothrophic action coupled with its participation in inflammation, metabolism and cardiovascular diseases (Amadio et al. 2017). The term 'triactome' clarifies the tight interaction between brain, immune system, and adipose tissue and their role in endorsing the development of cardiometabolic diseases (Singh et al. 2014). It is well known that the hypothalamic decline of BDNF controls energy equilibrium disturbing food consumption. BDNF haploinsufficiency or missense mutations in its receptor, TrkB (Singh et al. 2014) are linked to excessive eating and obesity. Cerebral BDNF insufficiency, on the other hand, catalyses a metabotropic loss which triggers obesity.

BDNF reacts functionally with the central anorexigenic and orexigenic signaling pathways. It has wide-ranging functions in adjusting energy homeostasis by controlling feeding, physical activity, and by adapting the metabolism of glucose in peripheral tissues. In addition to encouraging glucose transport and inducing mitochondrial genesis, BDNF fortifies and shields neurons against trauma and disease. By its action in the brain and periphery, BDNF augments insulin sensitivity and has a positive impact on insulin resistance (Rozanska et al. 2020). and the parasympathetic tone. Whereas, genetic factors, lifestyle, and chronic stress weaken BDNF signalling; and thus possibly substantiates the development of the metabolic syndrome. It's signaling enhances peripheral energy metabolism and cardiovascular function. Exercise and energy restraint cause BDNF signaling shortages and BDNF may subscribe to metabolic injury and allied diseases (Kellner et al. 2014).

A recent study of the feeding behaviour in mice proposed a concept of the central control of energy balance and obesity. This concept may have been moderated by the local dendritic alteration of a vital protein in neuronal flexibility brain-derived neurotrophic factor (Kellner et al. 2014).

Hypothalamic BDNF production is pivotal for normal energy balance and is a satiety indicator in the adult brain. The obesity of BDNF-depleted mice is solely the result of extreme calorie consumption (Fox et al. 2012). BDNF is also involved in apetite control and signaling pathways (Unger et al. 2007).

A recent noteworthy report of Rongrong et al. 2019, that the haploinsufficiency of the BDNF gene (chromosome 11p13) was accompanied by lower serum BDNF levels and childhood-onset obesity which were among individuals who were predisposed to Wager Syndrome. Multiple investigative studies on suicidal tendencies highlight the participation of BDNF in neurogenesis and synaptic plasticity. Those at the risk of committing suicide have lesser BDNF levels than those without a history of suicide attempts. Ostensibly this specific association to women's history of SA independent of probable psychiatric or demographic elements further enhances the role of BDNF as a promising biomarker for suicidal behaviour (Kudinova et al. 2019).

Furthermore, there is a possible link between BDNF and retinoic acid receptor in obesity emphasises the multifaceted pathogenecity of obesity coupled with behaviour and gender specific adiposity modifications (Brooke et al. 2010).

1.4 Leptin

Leptin is an anorexigenic hormone by its action in the hypothalamus. It participates significantly in controlling food consumption and feeding-linked performance. Leptin moderates an intricate system of many intercommunicating orexigenic and anorexigenic neuropeptides in the hypothalamus to lessen food consumption and enhance energy disbursement. Besides the arcuate nucleus, which is a key leptin sensitive centre, leptin-sensitive neurons in other hypothalamic nuclei, comprising the dorsomedial-, ventromedial-, paraventricular nucleus and the lateral hypothalamic regions are straight targets of leptin. Leptin also moderates hypothalamic neurons through the melanocortin system (Baver et al. 2014).

Leptin, which is present in excess, can adversely affect the male reproductive system. It is notable that regular leptin administration to rodents of normal weight induced comparable sperm aberrations. The key pathways responsible for these irregularities are yet to be revealed but these deleterious effects are ascribed to leptin-driven intensified oxidative stress as they are averted by simultaneous melatonin administration. Needless to highlight those studies on leptin and its effect on sperm function are decidedly pertinent in comprehending and handling male infertility, principally in overweight and obese men (Malik et al. 2019).

The pathogenesis of obesity is linked to leptin, uncoupling proteins and ion channels among many other factors. Leptin perse may control appetite via hypothalamic receptors, while uncoupling proteins existent in extracerebral cells may impact energy disbursement (Katharina and Jens 2017). In this context, congenital leptin deficiency is regarded as a cause of morbid obesity in early childhood.

Additionally, the process of 'uncoupling', where food breakdown and energy production are dissociated, is linked to various proteins that may be developed as useful in the treatment of obesity (Zaninovich 2005).

1.5 Vistafin (ElSaeed et al. 2019)

Visfatin is an adipokine secreted by visceral fat that has insulin-mimetic effects. The insulin-simulating function of visfatin/Nampt is still a contentious subject (Dalamaga and Christodoulatos 2017).

Recognised as a secreted growth factor, it augments pre B cell colony enhancing factor. Later, it was shown that this protein harbours an enzymatic nicotinamide phosphoribosyltransferase (Nampt) 2 activity, vital for the NAD synthesis, an indispensable cofactor for cell metabolism. Hence the terms visfatin, PBEF, and Nampt are synonymous.

Numerous tissues such as synovial fibroblasts, articular chondrocytes and monocytes generate visfatin which has features of typical cytokines (Luk et al. 2008). Thus it supports the notion that visfatin participates in different pathophysiological processes. Visfatin is also overexpressed in inflammatory diseases (Moschen et al. 2007). It is also up-regulated in cancer. Pro-inflammatory cytokines control visfatin expression; these pro-inflammatory cytokines comprise TNFα or IL-1β visfatin prompts chemotaxis, the generation of specific interleukins and TNF alpha by monocytes (Moschen et al. 2007).

Together, visfatin can be considered a crucial cytokine involved in a gamut of chronic inflammatory diseases and immune responses. Visfatin is also related to nicotinamide adenosine synthesis, with a role in degradation and synthesis of PGE2 prostaglandin E2 (Luk et al. 2008).

Besides, it has been shown that visfatin was unrelated to the regulation of insulin activity. Furthermore, Nampt activity is at best partly implicated in visfatin-induced PGE2 release in chondrocytes (Claire et al. 2012). The levels of visfatin were linked positively to insulin, FPG, HOMA-IR in those who were obese and negatively with adiponectin, which instead was positively linked to HDL-C and insulin. Leptin was only associated with BMI.

Studies support the role of visfatin in B-cell function (Han et al. 2020) and its possible implication in specific B-cell metabolic dysfunction. Seemingly, visfatin mediated leptin provoked UCP-1 mRNA expression and protein formation. As extraordinarily high concentrations of both visfatin and leptin caused a prominent lessening in UCP-1 protein levels, thus endorsing the concept that visfatin levels are elevated in obesity and that obese subjects have decreased BAT activity, conceivably through a decline in UCP-1 levels. Visfatin could exert differential control on various genes in brown adipose tissue. These findings have shed new light on the prospective working on visfatin in BAT (Dimitriadis et al. 2019).

1.6 Resistin

Resistin is a member of the protein-rich cysteines, in humans, different from rodents. It is mainly produced in macrophages, mononuclear cells and bone marrow cells (Fain et al. 2003). With pro-inflammatory cytokine activity, resistin is linked to obesity-related insulin resistance (Kusminski et al. 2005). Resistin is linked to the triad linking metabolism, inflammation and atherosclerosis and may well be a biomarker in these conditions (Muredach et al. 2005).

Mature adipocytes may or may not be the source of resistin. In obese diabetics resistin levels maybe increased (Degawa-Yamauchi et al. 2003). Resistin can be increased in inflammation and cardiac manifestations in the metabolic syndrome (Gerber et al. 2005; Reilly et al. 2005).

1.7 Ghrelin

Ghrelin is an endogenous hormone and a potent growth hormone secretagogue. A G protein-coupled receptor, GHS-R1a mediates its protective effects on cardiovascular tissue. The action of such growth hormone-releasing peptides and their binding sites may shed light on treatment strategies for cardiovascular diseases (Cao et al. 2006).

Circulating ghrelin is elevated in anorexia and cachexia and reduced in obesity. Both leptin and grehlin have crucial roles in maintaining metabolic equilibrium. Grehlin is much more than a "hunger hormone". It regulates energy homeostasis has cardio protective effects; has a role in bone metabolism.

Ghrelin and growth hormone secretagogue receptor impact many organs with useful and detrimental effects, there must be caution in avoiding side effects in developing therapies (Pradhan et al. 2013).

1.8 Chemerin

Chemerin is an adipose-tissue derived signalling molecule controls the formation of adipose tissue and the metabolism of adipocytes. Chemerin functions as the G protein-coupled CMKLR1 and participate in both the adaptive and the innate arms of immunity. Chimerin (and apelin, refer below) have key roles in the pathological processes of insulin resistance, glucose metabolism and obesity (Yu et al. 2012).

1.9 Apelin

Apelin has multiple phenotypic expressions and impacts several organs and tissues. Apelin increases energy consumption and insulin sensitivity, supporting it as a possible therapeutic target for obesity. Apelin and chimerin are linked to inflammation in obese patients suffering from Type 2 diabetics. Both these adipokines are linked to the development of events such as resistance to insulin, oxidative stress and inflammation (Yu et al. 2012).

To sum up, myokines, adipokines, BDNF, leptin, visfatin resistin, ghrelin, chimerin and apelin play important role in excess accumulation of lipids in adipose tissues.

References

Alireza AM, Mark W, Rachel H (2007) Review obesity and risk of colorectal cancer: A meta-analysis of 31 studies with 70,000 events. Cancer Epidemiol Biomarkers Prev 16(12): 2533-2547

Allen DL, Hittel DS, McPherron AC (2011) Expression and function of myostatin in obesity, diabetes, and exercise adaptation. Med Sci Sports Exerc 43: 1828-1835

Allott EH, Masko EM, Freedland SJ (2013) Obesity and prostate cancer: weighing the evidence. Eur Urol 63(5): 800-809

Amadio P, Colombo GI, Tarantino E, Gianellini S, Ieraci A, Brioschi M, Banfi C, Werba JP, Parolari A, Lee FS, Tremoli E (2017) BDNFVal66met polymorphism: a potential bridge between depression and thrombosis. Euro Heart J 38(18): 1426-1435

Baptiste R, Nadira YY, Paul DB, Alicia S (2019) TRPC5 ion channel permeation promotes weight gain in hypercholesterolaemic mice. Sci Rep 9: 773

Baver SB, Hope K, Guyot S, Bjørbaek C, Kaczorowski C, O'Connell KM (2014) Leptin modulates the intrinsic excitability of AgRP/NPY neurons in the arcuate nucleus of the hypothalamus. J Neurosci 34(16): 5486-5496

Beech DJ (2013) Characteristics of transient receptor potential canonical calcium-permeable channels and their relevance to vascular physiology and disease. Cric J 77: 570-579

Belizário JE, Oliveira CCF, Borges JP, Kashiabara JA, Edouard Vannier E (2016) Skeletal muscle wasting and renewal: a pivotal role of myokine IL-6. Springerplus. 2016; 5: 619

Brooke B, Kristie S, Stephen RW, Sun K, Santhosh G, Sarah H, Elsea R (2010) Haploinsufficiency causes reduced BDNF expression resulting in hyperphagia, obesity and altered fat distribution in mice and humans with no evidence of metabolic syndrome. Hum Mol Genet 19(20): 4026-4042

Cao JM, Ong H, Chen C (2006) Effects of ghrelin and synthetic GH secretagogues on the cardiovascular system Trends Endrocrinol Metab 17(1): 13-18

Chaldakov G (2011) The metabotrophic ngf and BDNF: An emerging concept. Arch Italiennes Biol 149: 257-263
Choukem SP, Tochie JN, Sibetcheu AT, Nansseu JR, Julian P. Hamilton-Shield JP (2020) Overweight/obesity and associated cardiovascular risk factors in sub-Saharan African children and adolescents: a scoping review. International Journal of Pediatric Endocrinology 2020: 6
Claire J, Martin H, Zvezdana M, Colette S, Emilie P, Francis B, Marjolaine G (2012) Pro-Inflammatory Actions of Visfatin/Nampt involve Regulation of Insulin Signaling Pathway and Nampt Enzymatic Activity* JBC Papers in Press. Published on March 7, 2012
Dalamaga M, Christodoulatos GS (2017) Visfatin, obesity, and cancer. In Adipocytokines, energy balance, and cancer 2017 (pp. 109-136). Springer, Cham
Danforth E, Horton ES, O'Connell M, Sims EA, Burger AG, Ingbar SH (1979) Dietary-induced alterations in thyroid hormone metabolism during over nutrition. J Clin Invest 64: 1336-1347
Debmalya S, Moutusi R (2016) Hypothyroidism and obesity: An intriguing link. Indian J Endocrinol Metab 20(4): 554
Degawa-Yamauchi M, Bovenkerk JE, Juliar BE, Watson W, Kerr K, Jones R (2003) Serum resistin (FIZZ3) protein is increased in obese humans. Clin Endocrinol Metab 88: 5452-5455
Dietrich P, Hellerbrand C (2014) Non-alcoholic fatty liver disease, obesity and the metabolic syndrome. Best Pract Res Clin Gastroenterol 28(4): 637-653
Dimitriadis GK, Adya R, Tan BK, Jones TA, Menon VS, Ramanjaneya M, Kaltsas G, Miras AD, Randeva HS (2019) Effects of visfatin on brown adipose tissue energy regulation using T37i cells. Cytokine 113: 248-255
Donath MY, Shoelson SE (2011) Type 2 diabetes as an inflammatory disease. Nat Rev Immunol 11: 98-107
Dorantes MTF , Lopez YED , Aguilar RG (2020) With Obesity and Their Impact on Neurodegenerative and Neurodevelopmental Diseases Front. Neurosci. 14: 863
Eckel RH, Krauss RM (1998) American Heart Association call to action: obesity as a major risk factor for coronary heart disease. AHA Nutrition Committee Circulation 97: 2099-2100
ElSaeed G, Mousa N, El-Mougy F et al (2019) Monogenic leptin deficiency in early childhood obesity https://doi.org/10.1111/ijpo.12574
Emily C, Dunford, Michael CR (2016) The metabolic implications of glucocorticoids in a high-fat diet setting and the counter-effects of exercise. Metabolites 6(4): 44
Fabio A, Ilaria P (2017) Measurement and clinical significance of biomarkers of oxidative stress in humans. Oxid Med Cell Longev 2017: 32
Fagot-Campagna A, Pettitt DJ, Engelgau MM, Burrows NR, Geiss LS, Valdez R, Beckles GLA, Saaddine J, Gregg EW, Williamson DF, K.M. Narayan V, (2000) Type 2 diabetes among North adolescents: An epidemiologic health perspective. The Journal of Pediatrics 136 (5):664-672
Fain JN, Cheema PS, Bahouth SW, Lloyd Hiler M (2003) Resistin release by human adipose tissue explants in primary culture. Biochem Biophys Res Commun 300: 674-678
Fardet L, Feve B (2014) Systemic glucocorticoid therapy: a review of its metabolic and cardiovascular adverse events. Drugs 74: 1731-1745
Fiore M, Chaldakov GN, Rancic G, Bełtowski J, Tunçel N (2014) *Aloe L.* an integrated view: Neuroadipocrinology of diabesity. Serbian J Exp Clin Res 15: 61-69
Fox EA, Biddinger JE, Jones KR, McAdams J, Worman A (2012) Mechanism of hyperphagia contributing to obesity in brain-derived neurotrophic factor knockout mice. Neuroscience 15: 229: 176-199
Freedman DS, Mei Z, Srinivasan SR, Berenson GS, Dietz WH (2007) Cardiovascular risk factors and excess adiposity among overweight children and adolescents: the Bogalusa Heart Study. J Pediatr 150: 12-17
Furukawa S, Fujita T, Shimabukuro M, Iwaki M, Yamada Y, Nakajima Y, Nakayama O, Makishima M, Matsuda M, Shimomura I (2004) Increased oxidative stress in obesity and its impact on metabolic syndrome. J Clin Invest 114: 1752-1761

Gerber M, Boettner A, Seidel B, Lammert A, Bar J, Schuster E (2005) Serum resistin levels of obese and lean children and adolescents: biochemical analysis and clinical relevance. J Clin Endocrinol Metab 90: 4503-4509

Han DF, Yang Li, Xu HY, Li RH, Ding Zhao D, (2020) An Update on the Emerging Role of Visfatin in the Pathogenesis of Osteoarthritis and Pharmacological Intervention. Evidence-Based Complementary and Alternative Medicine 1-7

Jana V, van Vliet-Ostaptchouk, Harold S, Vasiliki L (2012) Gene–lifestyle interactions in obesity. Curr Nutr Rep 1(3): 184-196

Jorge J. Castillo, Randall R. Ingham, John L. Reagan, Michael Furman, Samir Dalia, Joanna Mitri, (2014) Obesity Is Associated With Increased Relative Risk of Diffuse Large B-Cell Lymphoma: A Meta-Analysis of Observational Studies. Clinical Lymphoma Myeloma and Leukemia 14 (2):122-130

Katharina T, Jens CB (2017) Hypothalamic circuits regulating appetite and energy homeostasis: pathways to obesity Dis Model Mech 10(6): 679-689

Keaney JF, Larson MG, Vasan RS, Wilson PW, Lipinska I, Corey D, Massaro JM, Sutherland P, Vita JA, Benjamin EJ (2003) Obesity and systemic oxidative stress: clinical correlates of oxidative stress in the Framingham Study. Arterioscler Thromb Vasc Biol 23: 434-439

Kellner Y, Gödecke N, Dierkes T, Thieme N, Zagrebelsky M, Korte M (2014) The BDNF effects on dendritic spines of mature hippocampal neurons depend on neuronal activity. Front Synaptic Neurosci 6:5

Keys A, Fidanza F, Karvonen MJ, Kimura N, Taylor HL (1972) Indices of relative weight and obesity. J Chron Dis 25(6): 329-343

Khaodhiar L, McCowen KC, Blackburn GL(1999) Obesity and its comorbid conditions. Clinical Cornerstone 2 (3):17-31

Kudinova AY, Deak T, Deak MM, Gibb BE (2019) Circulating levels of brain-derived neurotrophic factor and history of suicide attempts in women. Suicide Life Threat Behav 49(1): 54-63

Kusminski CM, McTernan PG, Kumar S (2005) Role of resistin in obesity, insulin resistance and Type II diabetes. Clin Sci (Lond) 109(3): 243-256

Kutty MK , Osman MT Obesity and Cancer Current Concepts 2013 Edition: 1stPublisher: LAP LAMBERT Academic Publishing, GermanyISBN: 978-3-659-44993-2

Luk T, Malam Z, Marshall JC (2008) Pre-B-cell colony-enhancing factor (PBEF)/visfatin: a novel mediator of innate immunity. J Leukocyte Biol 83: 804-816

Malik IA, Durairajanayagam D, Singh HJ (2019) Leptin and its actions on reproduction in males. Asian J Androl 21(3): 296-299

Maria D (2013) Interplay of adipokines and myokines in cancer pathophysiology: Emerging therapeutic implications. World J Exp Med 3(3): 26-33

Mario SP, Guadalupe EG, Otilia PP, Araceli ME, Mario GH,3 Héctor BO, O Eyerahi BF, Arturo CP, Veronica ZC, Ethel GL, Gabriela GP, José Alfredo HP, Claudine I (2018) Key clinical factors predicting adipokine and oxidative stress marker concentrations among normal, overweight and obese pregnant women using artificial neural networks. Int J Mol Sci 19(1): 86

Marshall A. Lichtman, (2010) Obesity and the Risk for a Hematological Malignancy: Leukemia, Lymphoma, or Myeloma. The Oncologist 15 (10):1083-1101

Michaela AO, Rosemarie ES, Karen HL (2016) Addressing the role of obesity in endometrial cancer risk, prevention, and treatment. J Clin Oncol 34(35): 4225-4230

Monteiro R, Azevedo I (2010) Chronic inflammation in obesity and the metabolic syndrome. Mediat Inflamm 2010: 289645

Moschen AR, Kaser A, Enrich B, Mosheimer B, Theurl M, Niederegger H, Tilg H (2007) Visfatin, an adipocytokine with proinflammatory and immunomodulating properties. J Immunol 178: 1748-1758

Muredach PR, Michael L, Megan LW, Anand R, Mitchell AL, Daniel JR (2005) Resistin is an inflammatory marker of atherosclerosis in humans. Circulation 111: 932-939

Nakahashi T, Fujimura H, Altar CA, Li J, Kambayashi J, Tandon NN, Sun B (2000) Vascular endothelial cells synthesize and secrete brain-derived neurotrophic factor. FEBS Lett 470: 113-117

Ng M, Fleming T, Robinson M, Thomson B, Graetz N, Margono C (2014) Global, regional, and national prevalence of overweight and obesity in children and adults during 1980-2013: a systematic analysis for the Global Burden of Disease Study 2013. Lancet 384: 766-781

Nuttall, Frank Q (2015) Body Mass Index: Obesity, BMI, and Health A Critical Review. Nutr Today 50(3): 117-128

Pradhan G, Samson SL, Sun Y (2013) Grehlin: much more than a hunger hormone. Clin Opin Nutr Metab Care 16(6): 619-624

Prasenjit M, Sushil KJ (2015) Obesity, oxidative stress, adipose tissue dysfunction, and the associated health risks: causes and therapeutic strategies. Metab Syndr Relat Disord 13(10): 423-444

Ramírez A, Vázquez-Sánchez AY, Carrión-Robalino N, Camacho J (2016) Ion channels and oxidative stress as a potential link for the diagnosis or treatment of liver diseases. Oxid Med Cell Longev 2016

Reilly MP, Lehrke M, Wolfe ML, Rohatgi A, Lazar MA, Rader DJ (2005) Resistin is an inflammatory marker of atherosclerosis in humans. Circulation 111: 932-939

Renchan AG, Tyson M, Egger M, Heller RF, Zwahlen M (2008) Body-mass index and incidence of cancer: a systematic review and meta-analysis of prospective observational studies. Lancet 371: 569-578

Rongrong H, Zeyue L, Nannan S, Shu L, Lanlan L, Yan S, Jianbo X, and Qi X (2019) BDNF alleviates neuroinflammation in the hippocampus of type 1 diabetic mice via blocking the aberrant HMGB1/RAGE/NF-κB pathway. Aging Dis 10(3): 611-625

Rosanna P (2019) Exercise-induced myokines with therapeutic potential for muscle wasting. Front Physiol 10: 287

Rozanska O, Uruska A, Ziolkiewicz DZ (2020) Brain-Derived Neurotrophic Factor and Diabetes. International Journal of Molecular Sciences 21 (3):841

Schwenger KJ, Bolzon CM, Li C, Allard JP (2019) Non-alcoholic fatty liver disease and obesity: the role of the gut bacteria. Eur J Nutr 58(5): 1771-1784

Singh RB, Takahashi T, Tokunaga M, Wilczynska A, Kim CJ, Meester FD, Handjieva-Darlenska T, Cheema SK, Wilson DW, Milovanovic B, Fedacko J (2014) Effect of brain derived neurotrophic factor, in relation to diet and lifestyle factors, for prevention of neuropsychiatric and vascular diseases and diabetes. Open Nutraceuticals J 7(1)

Unger TJ, Calderon GA, Bradley LC, Sena-Esteves M, Rios M (2007) Selective deletion of BDNF in the ventromedial and dorsomedial hypothalamus of adult mice results in hyperphagic behavior and obesity. J Neuroscience 27(52): 14265-14274

van Staa TP, Cooper C, Leufkens HG, Lammers JW, Suissa S (2003) The use of inhaled corticosteroids in the United Kingdom and the Netherlands. Respir Med 97: 578-585

Vasconcelos LH, Souza IL, Pinheiro LS, Silva BA (2016) Ion channels in obesity: pathophysiology and potential therapeutic targets. Front Pharmacol 7: 58

WHO (1995) Physical Status: The Use and Interpretation of Anthropometry: Report of a World Health Organization (WHO) Expert Committee. Geneva, Switzerland: World Health Organization

Yu S, Zhang Y, Li M-Z, Xu H (2012) Chemerin and apelin are positively correlated with inflammation in obese type 2 diabetic patients. Chin Med J 125(19): 3440-3444

Zaninovich AA (2005) Role of uncoupling proteins UCP1, UCP2 and UCP3 in energy balance, type 2 diabetes and obesity. Synergism with the thyroid. Medicina (B Aires) 65(2): 163-169

Zhiyong C, Ping Z, Zhiqing H, Jing Y, Chun L, Yusheng R, Zonggui W (2016) Obstructive sleep apnea combined dyslipidemia render additive effect on increasing atherosclerotic cardiovascular diseases prevalence. Lipids Health Dis 15: 98

The Economics of Obesity: A Public Policy Perspective

2

Vishnu Raj Kumar and Prameela Kannan Kutty

Abstract

While a large segment of this book has been devoted to the medical analysis of the causes and consequences of obesity, this chapter will analyse the subject using the tools of a social scientist. One must recognize, that obesity is a multifaceted disease that affects not only the health of a person but also his or her socioeconomic standing, as well as the productive potential of the wider economy. As such, any serious attempt to treat obesity at the communal level should not only analyse the subject from a purely medical and scientific perspective, but should also give serious consideration to how proper public policy administration can be effectively used to lower obesity rates in society and to encourage a healthier lifestyle. This chapter will focus (1) on the socioeconomic impact of obesity, (2) evaluation of several public policy solutions to tackle obesity in society, (3) and socioeconomic consequences of obesity by analysing its effect on both the individual and the wider economy; As such both a micro and macroeconomic analysis of obesity will be presented. Furthermore, this chapter will evaluate how policymakers can tackle obesity through the interplay of policy incentives, interest groups and economic institutions. The authors ultimately believe that there is no one best solution to tackle obesity. Like any serious public health problem, tackling obesity requires the collective and coordinated efforts of multiple parties that should employ both precise medical solutions as well as the right policy tools.

V. R. Kumar
London School of Economics, Cambridge, UK

P. K. Kutty (✉)
National University of Defence, Kuala Lumpur, Malaysia

M. K. Kutty, A. Elengoe (eds.), *Obesity and its Impact on Health*,
https://doi.org/10.1007/978-981-33-6408-0_2

Keywords

Economics · Public policy · Obesity

2.1 The Consequences of Obesity

From a socioeconomic perspective, obesity impacts both the individuals, affecting his or her productive and earning capacity as well as the wider economy. This segment will review previous socioeconomic literature to see exactly how obesity can impact society in this way.

2.2 Obesity and the Individual

This segment is a microeconomic evaluation of obesity and reviews its impact on the individual. While there are many consequences of obesity on the individual, this segment examines its impact on an individual's employment and direct earning potential. It also looks at the psychosocial consequences of obesity on an individual and how this in turn impacts their productivity.

2.3 Income and Employment

It is widely known that obese individuals earn lower levels of income compared to individuals who are not obese. This is because there is evidence that obese people are less likely to find employment (Flint and Snook 2014) and when they do find employment, obese individuals are often associated with lower levels of pay compared to non-obese individuals (Averett 2014). For the analysis of obesity and its linkage with employment discrimination, Flint et al. employ a robust statistical study to examine a sample of 181 individuals from wide-ranging types of employment broadly categorized into sedentary, standing, manual and heavy manual occupations. Their results tell us that across all four categories, the weight status of individuals impact judgments on their suitability for work and employment. Furthermore, the study highlights that the more physically demanding a job, the less favourable judgments are towards obese individuals. Ball et al. (2002) also find evidence that tells us that obesity and high BMI are linked with jobs that are of lower economic status than non-obese individuals. Flint and Snook (2014) research show that obesity is linked to discrimination in the hiring stage where obese individuals are assessed to have less leadership potential, are less likely to be employed and are likely to be less successful than their non-obese counterparts. Most interestingly, it is found that obese individuals are subject to negative stereotypes by hiring managers even before they are interviewed and in many cases, this prevents them from the very initial interview as part of the hiring process of firms (Agerström and Rooth 2011).

When it comes to obesity and income levels, various studies conducted in the US and European countries find negative and statistically significant effects of obesity on earnings (Averett 2014). This, in turn, could potentially be a result of Levine and Schweitzer (2015) findings which show that obese individuals tend to be associated with lower levels of competence and hence lower productivity levels compared to non-obese individuals. Similarly, Schulte et al. (2007) study finds evidence that obese individuals receive reduced initial salaries, are perceived to be less qualified despite similar educational and professional backgrounds and tend to work longer hours than non-obese individuals. There is substantial evidence to show that gender confounds differences in income and employment opportunities between obese and non-obese persons. For instance, Roehling (1999) finds that obese women are 300% more likely to voice discrimination in employment compared to obese men. Frederickson and Roberts (1997) argue that this is because unlike men, women are evaluated on a basis of beauty standards and sexual attractiveness which makes obesity far more consequential for employment in women as opposed to men. This further discrimination in the employment of obese women is reinforced by Bartels and Nordstrom (2013) study of the role of gender and obesity in employment.

All this tells us that obesity has serious and tangible consequences on the income-generating capacity of an individual as they are less likely to find employment and when they do, they are positively discriminated against due to existing inherent biases prevalent among managers as well as the lower productivity levels that tend to be widely associated with individuals that are obese.

2.4 Obesity and Its Psychosocial Impact

It is well known that obese individuals suffer from various social and cultural stigmas that have direct and indirect consequences on their socioeconomic standing. Franklin et al. (2006) finds that obese adolescents tend to have lower levels of self-esteem and self-worth when compared to non-obese adolescents. Such findings were more prominent among females. Although not all obese persons suffer from this, the majority of obese people are victims of lower self-worth and as a result tend to shy away from competition in academia, sports, employment as well as general public engagement. As such, one could deduce that obese persons are generally less likely to be strivers in society as opposed to non-obese individuals. Such individuals are likely to be less productive in society and will likely have stunted ambition. This confounds the existing disadvantages that obesity already has in the labour market as described in the above section. As a result, this leads to a negative vicious cycle that obesity has on the economic and financial standing of individuals. Indeed Gortmaker et al. (1993) finds that obese adolescents are associated with lower levels of eventual household income as well as lower eventual marital rates compared to non-obese individuals with these differences being more prominent among women compared to men. It is important to note however that in their research, Gortmaker et al. (1993) finds no evidence of obesity on self-esteem. The link that obesity has on marriage is of particular importance. For instance, in societies and cultures in developing parts of the world where arranged marriages and marrying one's children off are seen to relieve a family's existing economic burden (Kadam 2015). As such, lower marital rates among obese individuals can be seen to add to the economic burden of households in rural communities.

It is therefore evident that through its effect on self-esteem, the psycho-social consequences of obesity actively discourage competitive and productive individual behavior. This has direct consequences for an individual's social and financial standing.

We conclude this segment by arguing that obesity exogenously and endogenously negatively impact an individual's social and economic standing. Exogenous consequences stem from the fact that obese individuals are discriminated against by employers and thus face lower levels of pay and opportunities in the workplace as a result of intrinsic biases. Endogenous consequences arise from the lower productivity levels from obese individuals as well as reduced self-esteem which leads to uncompetitive behavior and suppressed ambition and thus reduced earning potential from obese persons.

2.5 Obesity and the Aggregate Economy

When the healthcare costs, foregone opportunities and reduced productivity levels associated with obese individuals are aggregated, obesity poses a serious cost to the wider economy. Globally, the aggregate economic costs of obesity amounts to ~USD 2 trillion annually or ~2.8% of global GDP (Nath 2019). In fact, coupled with armed violence and smoking, obesity is considered to be within the top three gravest social burdens generated by humans (Nath 2019). This segment analyses the consequences of obesity on the aggregate economy by examining the direct economic burden of the disease as well as the indirect costs of obesity and its overall impact on the social and political economy we live in today through the role that obesity has on income inequality.

2.5.1 Direct and Indirect Costs

In terms of its direct economic costs, the prevalence of obesity in society acts as a strain on public finances, particularly the healthcare costs of a nation. For instance in the US even in the 1990s, between 4.3% (Allison et al. 1999) and 7.8% (Colditz 1992) of total healthcare costs can be directly attributed to obesity and its related diseases. For other countries inclusive of France, New Zealand, Australia and Canada, this figure varies between 1% and 4% (Bagust et al. 1999; Müller-Riemenschneider et al. 2008). Despite being lower than the US, the figure still remains absolutely critical as it relates to a government's fiscal responsibilities. Additionally, the reduced economic productivity (Dee et al. 2014) that stems from obese societies also significantly affects a country's economic potential. In a study conducted by Kudel et al. (2018), obesity was found to impact the productivity and thus the output of different occupations and various economic sectors differently. Kudel et al. (2018) found that the construction, arts and hospitality sectors suffer the most significant productivity losses when employees are obese. Hence, it can be deduced from Kudel et al. (2018) research that the more important such sectors are for a national economy, the more serious the impact of obesity on total national economic output and potential. In fact, Nath (2019) estimates that productivity losses stemming from obesity can cost employers about $506 per worker in sick days and absenteeism. In terms of indirect costs, obesity presents foregone opportunities for an economy. Public expenditure for obesity related medical treatment (in countries where there is a state sponsored healthcare program) could have been spent on more productive investments in the economy. For instance, had that money been invested in the development of roads and schools, this would add to the productive potential of an economy. Economists use the term "opportunity costs" to describe such foregone opportunities (Dee et al. 2014). Similarly, premature mortality and missed days from work are seen as indirect costs of obesity (Dee et al. 2014) on society which affects the productive potential of an economy and its outlook for long term growth. As previous studies have shown, the ultimate economic impact of obesity is inherently complicated to measure (Kudel et al. 2018) and in many cases they are often under measures. One reason for this is that it is intrinsically difficult to measure the full extent of obesity the economic, earning potential and a society's drive to take risks and succeed.

As such, one can see that as a public health pandemic, obesity has serious and tangible consequences on the economic development of societies. In order for countries to develop sustainably, obesity is a key public policy issue that needs attention.

2.5.2 Inequality

The relationship between obesity and inequality is one that is worth examining further. On the one hand, we find that in developed countries, obesity tends to be more prevalent among individuals and households that are of lower income levels (Bhurosy and Jeewon 2014). In general, it is believed that those of a higher income and socioeconomic status are able to consume a diet that is of better quality which includes fresh produce and proteins that tend to be more expensive (Sharkey et al. 2011). Moreover, individuals that are richer are also able to gain access to services such as gym memberships and personal trainers as well as be more conscious about the latest dietary and lifestyle trends which prompt a healthier lifestyle. Conversely, poorer segments of society tend to consume an energy-dense diet that is high in its calorie, trans fat and cheap vegetable oil content. As such, one can deduce that lower levels of income can be a driver for obesity in developed societies. (Mclaren 2007). Interestingly, in less developed countries, these observations are reversed. Obesity appears to be more prevalent among the richer segments of society (Kumanyika et al. 2002). In fact, Kumanyika et al. (2002) show a positive relationship between increasing socioeconomic status and the prevalence of obesity in lower income countries. One potential explanation for this disparity in the prevalence of obesity between high and low income countries, could be the fact that in lower income countries, higher socioeconomic status leads to greater access to consuming high calorie food and lower physical labour whereas, in higher income countries, high socioeconomic status leads to a greater awareness for a healthier and more balanced lifestyle (Houle 2013). All this implies that while improvements in health come as a result of economic development, complications that arise from malnutrition are substituted by problems of overconsumption which asymmetrically affect varying socioeconomic groups (Houle 2013; US Centre for Disease Control and Prevention 2012).

So far, this segment has shown how socioeconomic status and income inequality levels can contribute to obesity depending on a country's level of development. However, it is also worth noting that obesity in turn can be a driver for income inequality across countries. This happens as obese individuals are more likely to end up in jobs that pay less compared to non-obese persons due to factors such as labour-market discrimination, self-esteem issues and public biases (Kim and Knesebeck 2018) as explored in the earlier part of this chapter. As a result, countries with growing rates of obesity can expect to see this have an impact on its Gini coefficients. From a total of 21 studies on obesity in the UK, US and Canada, Kim and Knesebeck (2018) find evidence that obesity can be a driver for income differentials in a third of the studies.

As such, the relationship between obesity and inequality levels in society remain intrinsically complex. Obesity can be a contributor and symptom of income inequality in society. In a world where income inequality has emerged as a central theme in global politics, policy makers would do well to closely integrate efforts to tackle this public health problem in society.

2.6 Tackling Obesity

In this segment we evaluate the process of tackling obesity in society. In this we look at how policy makers can use incentives to nudge society to live a healthier lifestyle. We also look at the role of interest groups as well as the challenges and opportunities they present to policy makers when it comes to tackling obesity. Finally, we briefly evaluate existing economic legislation that has been enacted to tackle this public health problem.

2.6.1 Policy Incentives

Here, using the literature from behavioral economics, we look at how governments can use policies to influence consumption decisions to tackle obesity in society. Essentially, behavioral economists think of individuals of having a double persona; a short-sighted self that make almost immediate consumption decisions and a far-sighted self that make live with the health consequences of decisions made

(Faulkner et al. 2011). Sometimes there is a conflict in the actions of these two selves. This can occur when the immediate consumption decisions of an individual fails to take into account the welfare effects of their future self and thus may consume unhealthy food (Faulkner et al. 2011). It is therefore the role of the policy maker to incentivize the individual to act in a way that is beneficial to his welfare in the future.

This segment evaluates the fiscal policy tools of taxes and subsidies that are available to policy makers. We look at how they may work as well as analyse their wider economic, social and political consequences on society.

2.6.2 Taxes

The use of taxes on high calorie and sugary foods as a means to tackle obesity continues to gain prominence and remains a subject of significant discussion among economists and public health experts. The central economic argument for such a tax is that the associated price increase that such a tax would have on food is likely to divert consumption away from fatty food towards healthier alternatives (Maniadakis et al. 2013). This in turn is expected to reduce obesity and encourage a healthier diet. Such a policy has shown some progress. Bascunan and Cuadrado (2017) have studied the impact of such a sugar tax on sweet drinks in Chile that was imposed in 2014. Their analysis shows a decrease in the prevalence of obesity ranging from 0.99% to 2.4% among the sample in their study group. Similarly a study by the University of California Berkeley (Falbe et al. 2016) reveals a 21% drop in the consumption of sugary non-alcoholic beverages and a 29% increase in the consumption of water just months after an introduction of a soda tax in Berkeley. This provides further evidence that such a policy measure can actively discourage the consumption of unhealthy food and encourage the consumption of healthier alternatives which could yield positive results in tackling the prevalence of obesity and its associated comorbidities.

It is worth also considering the arguments against such a policy to curb obesity. Firstly, interventions of such a kind by policy makers are seen to violate the sovereignty of consumers which may not be a politically viable economic policy for capitalist societies (Maniadakis et al. 2013). Secondly, such taxes are consumption taxes and as such they are regressive in nature (Maniadakis et al. 2013) which may further widen existing inequality trends that are widely prevalent in society. Third, consumption behavior is inherently complex and difficult to predict and as such while some consumers may switch to consuming healthier alternatives, other consumers may switch to consume alternatives which are even more detrimental to health, thus exacerbating obesity related diseases (Fletcher et al. 2010). Alternatively, if consumers are addicted to the foods being taxed, some consumers may decrease their consumption of healthier alternatives such as fruits and vegetables just so as to maintain their level of expenditure on the taxed items (Maniadakis et al. 2013), hence completely undermining the very purpose of such a tax in the first place.

While, targeted taxes can be a suitable tool to create the right consumption incentives to tackle obesity, their application should be exercised with the right degree of caution and precision. One suggestion as proposed by Maniadakis et al. (2013) would be to use broad instead of narrow taxes. This would help prevent consumers from switching to unhealthier alternatives and would nudge them in the direction of consuming healthier alternatives. However, given the social, economic and political dimensions of complexities that underpin tax policies in many countries, the viability of introducing a tax that adequately tackles obesity remains slim and it should instead be combined with other public policy proposals.

2.6.3 Government Expenditure: Subsidies

As far as public finance is concerned, subsidies are essentially the inverse of taxation. With regards to obesity, this policy proposal looks at subsidizing healthier alternatives that incentivize healthier consumption to tackle obesity in society. In their research, Faulkner et al. (2011) find that weight outcomes are responsive to prices of food and beverages. As such, they find that policies such as subsidies that reduce the price of healthier food and thus incentivize healthy consumption are positive in the fight against obesity. Similarly, McCarthy (2016) reinforces the view that consumers respond to reduced food prices. For instance, in his research he finds that in the US, adults that ate the most foods that benefited from agricultural subsidies were at a higher risk of obesity. Such food generally was richer in saturated fats, refined grains, sugar and salt. All this tells us that, price can be a key factor in influencing dietary decisions and as far as subsidies are concerned, governments can use such fiscal tools in a way that positively encourages healthier food intake.

It is however worth noting that just like taxes, subsidies interfere with the market mechanism which can distort consumer choice, reduce welfare and result in deadweight losses in the market (Faulkner et al. 2011). Additionally, implementing a subsidy requires governments to incur labour administrative costs as well as budgetary costs (Faulkner et al. 2011). Finally, as mentioned earlier, consumption decisions are by nature hard to understand and at the end of the day, subsidies may not end up producing the desired results that it was intended to. Given the fiscal consequences of a subsidy to the wider economy, such a decision needs to be exercised with caution alongside other public policy decisions.

2.6.4 Interest Groups

This section looks at the role of interest groups in tackling obesity in society. In doing so, the politics of obesity is analysed. This includes a look at the resistance that policy makers face from vested interest groups as well as the effectiveness of campaigns and coordinated efforts to counter this resistance and to push for reforms in this space.

2.6.5 Focused Interest Groups and the Effectiveness of Campaigns

Like all serious public policy initiatives, any effort to eradicate and tackle obesity in any given society will be met by both vehement opposition against it and strong support for it. In the general literature of political science and economics, it is widely known that any actor that derives direct or indirect benefit from a set of existing institutions and polices, will always support the status quo and therefore "reinforce a path-dependent process" (Banchoff 2005). In this regard, it is often believed that large food and beverage corporations, sugar companies and parties that stand to benefit from overconsumption of their products, are often the ones that lobby against serious reform that could eat into their profits (Kersh 2009). In many ways the opposition in the obesity debate is parallel to the resistance faced by policymakers from the tobacco and alcohol lobbies (Kersh 2009). This is very often done by framing the debate in a way that is politically contentious that ultimately aims to discourage serious legislation in that space (Kersh and Morone 2002). For instance, as it relates to obesity, interest groups have framed the debate around a narrative of "personal responsibility" (Kersh 2009). This narrative essentially claims that eating the right food and leading a healthy lifestyle is the "personal responsibility" of every individual. As such, in a capitalist society that values freedom of choice like the United States, overt encroachment by the state to regulate consumption, can be seen as undermining and even violating this individual liberty. Prior practices such as helmet-wearing, alcohol consumption and smoking have also previously been framed in this language of "individual rights" (Goodin 1990). The ultimate aim is to prevent concrete legislation and to preserve the financial and political interests of those that benefit from the status quo (Kersh 2009). However, this isn't to say that such resistance can't be countered. Indeed, there have been serious efforts to question the ethical and medical implications of framing issues using the "personal responsibility" method (Guttman and Ressler 2001). There have also been concerted and coordinated lobbying efforts from groups such as the American Diabetes Association, American Dietetic Association, and American Heart Association in the US (Kersh 2009), as well as similar associations in other countries that have pushed reform in the right direction for the fight against obesity. In fact in the US, despite the resistance, there continues to be small and effective groups of coordinated policy makers in Congress that push for positive reform in this area. Even former President Bill Clinton has emerged as a positive voice in this debate (Kersh 2009). Such actions from various groups continue to catalyse campaigns and action for actively tackling obesity in society. All this have translated into tangible action such as mandating food labelling in 1990 (Kersh and Morone 2002) in the United States.

The purpose of the above segment is to highlight the political complexities of bringing tangible reform to tackle the obesity pandemic in society. Ultimately, bringing about policy changes requires time and coordinated effort to unshackle vested interest groups. Past events tell us that such effort is possible, but it remains a painstaking and time-consuming endeavor which is worth undertaking.

2.6.6 Economic Institutions/Legislation

So far, we have evaluated how the right incentives can be organized to tackle obesity. We have also looked at the challenges and opportunities faced by policy makers in their efforts to tackle obesity in society. In this final segment, we will briefly evaluate the effectiveness of existing economic institutions and legislation that have been introduced to combat obesity in society. As briefly alluded to earlier, one significant and concrete legislative achievement has been the introduction of food labelling in the US by congress which identifies the content of food, thus helping consumers make consumption decisions (Kersh and Morone 2002). Although there remains much more work to be done as it relates to the content of such labels, such an introduction is a step forward in the public policy process of tackling obesity in some societies. Bonsmann and Wills (2012) argue that mandatory labelling by the EU for food products is integral to allowing consumers to make the right decisions about their health. They however assert that labels alone without concerted public campaigns to raise the awareness of the importance of healthy eating are unlikely to be sufficient to effectively reduce obesity rates in societies. Apart from labelling, government officials can use the platforms and institutions of government to publicize nutritional warnings in an attempt to induce healthier lifestyles. In the US for instance government sponsored campaigns that linked cancer and heart disease to the consumption of saturated fats saw a statistically significant decline in saturated and other fats between 1977 and 1985 (Ippolito and Mathio 1995). Despite vested interests protecting the status quo, it is found that constant publicity campaigns have resulted in a more concrete regulatory regime in some countries that actively monitor and regulate food intake of certain pockets of society (Kersh and Morone 2002). In the US for example, we see the creation of independent regulatory agencies such as and programs such as the National School Lunch Program, the creation of several units within the US Department of Health and Services including the FDA and the National Institutes of Health among others that put major emphasis on issuing guidelines on food and beverage intake and recommendation for different societal groups (Kersh and Morone 2002). Despite their shortcomings, the fight against obesity is better with their efforts rather than without.

2.7 Conclusion

This chapter has strived to analyse the economics of obesity from a public policy perspective. The authors continue to believe that although obesity is fundamentally a medical problem, an effective and serious approach to treat it requires attacking the subject from a medical as well as a public policy perspective. The author first began this chapter by looking at the tangible economic consequences of obesity on an individual and then on the national economy. The author subsequently looked at policy responses to tackle obesity in light of organizing the right incentives for consumers, the role and opposition of vested interest groups as well the potential of economic and legislative institutions. What remains clear from our evaluation is that tackling this disease requires a multi-dimensional and multi-perspective approach and understanding of the science of obesity as well as of the political economy of the disease. It is the authors' hope that this chapter has been able to shed some light on the latter as we all continue to play our role in tackling this pandemic.

References

Agerström J, Rooth D (2011) The role of automatic obesity stereotypes in real hiring discrimination. J Appl Psychol 96: 790-805

Allison D, Zannolli R, Narayan KV (1999) The direct health care costs of obesity in the United States. Am J Public Health 89(8): 1194-1199

Averett S (2014) Obesity and labor market outcomes. IZA World Labor 2014

Bagust A, Lynne-Roberts B, Haycox AR, Barrow S (1999) The additional cost of obesity to the health service and the potential for resource savings from effective interventions. Eur J Public Health 9(4), 258-264

Ball K, Mishra G, Crawford D (2002) Which aspects of socioeconomic status are related to obesity among men and women? Int J Obes 26: 559-565

Banchoff W (2005) Path dependence and value-driven issues: The comparative politics of stem-cell research. World Polit 57(1): 200-230

Bartels LK, Nordstrom CR (2013) "Too big to hire: factors impacting weight discrimination," Management Research Review, Emerald Group Publishing, 36(9), 868-881

Bascuñán J, Cuadrado C (2017) Effectiveness of sugar-sweetened beverages taxes to reduce obesity: evidence brief for policy. Medwave 17(8): e7054

Bhurosy T, Jeewon R (2014) Overweight and obesity epidemic in developing countries: a problem with diet, physical activity, or socioeconomic status? Sci World J 2014: 964236

Bonsmann SSG, Wills JM (2012) Nutrition Labeling to Prevent Obesity: Reviewing the Evidence from Europe. Current Obesity Reports 1 (3):134-140

Colditz G (1992) Economic costs of obesity. Am J Clin Nutr 55(2): 503S–507S

Dee A, Kearns K, O'Neill C, Sharp L, Staines A, O'Dwyer V, Fitzgerald S, Perry IJ (2014) The direct and indirect costs of both overweight and obesity: a systematic review. BMC Res Notes. 16;7:242

Falbe J, Thompson HR, Becker CM, Rojas N, McCulloch CE, Madsen KA (2016) Impact of the berkeley excise tax on sugar-sweetened beverage consumption. Am J Public Health 106 (10) 1865-1871

Faulkner GE, Grootendorst P, Nguyen VH, Andreyeva T, Arbour-Nicitopoulos K, Auld MC (2011) Economic instruments for obesity prevention: results of a scoping review and modified Delphi survey. Int J Behav Nutr Phys Act 8: 109

Fletcher JM, Frisvold D, Tefft N, (2010) Can soft drink taxes reduce population weight ? Contemporary Economic Policy 28 (1):23-35

Flint SW, Snook J (2014) Obesity and Discrimination: the Next 'Big Issue'? Int J Discrimin Law 14: 183-193

Franklin J, Denyer, Steinbeck KS, Caterson ID, Hill AJ, (2006) Obesity and Risk of Low Self-esteem: A Statewide Survey of Australian Children. Pedaitrics 118 (6):2481-2487

Fredrickson BL, Roberts TA (1997) Objectification theory: Toward understanding women's lived experiences and mental health risks. Psychol Women Q 21: 173-206

Goodin RE (1990) No Smoking: The Ethical Issues. Chicago: University of Chicago Press.

Gortmaker SL, Must A, Perrin JM, Sobol AM, Dietz WH (1993) Social and economic consequences of overweight in adolescence and young adulthood. N Engl J Med 329(14): 1008-1012

Guttman N, Ressler WH (2001) On being responsible: ethical issues in appeals to personal responsibility in health campaigns J Health Commun 6(1): 117-136
Houle B (2013) How obesity relates to socioeconomic status, how obesity relates to socioeconomic status
Ippolito PM, Mathios AD (1995) Information and advertising: The case of fat consumption in the United States. Am Econ Rev 85(2): 91-95
Kadam, T (2015) "The Impact of Arranged Marital Customs on Women's Autonomy in Rural India". Student research. Paper 29
Kersh R (2009) The Politics of Obesity: A Current Assessment and Look Ahead. Milbank Quarterly 87 (1):295-316
Kersh R, Morone J (2002) The politics of obesity: Seven steps to government action. Health Aff 21 (6): 162-175
Kim TJ, von dem Knesebeck O (2018) Income and obesity: what is the direction of the relationship? A systematic review and meta-analysis. BMJ Open 8: e019862
Kudel I, Huang JC, Ganguly R (2018) Impact of Obesity on Work Productivity in Different US Occupations. Journal of Occupational and Environmental Medicine 60 (1):6-11
Kumanyika S, Jeffery RW, Morabia A, Ritenbaugh C, Antipatis VJ (2002) Obesity prevention: the case for action. Int J Obes 26(3): 425-436
Levine EE, Schweitzer ME (2015) The affective and interpersonal consequences of obesity. Organ Behav Hum Decis Process 127: 66-84
Maniadakis N, Kapaki V, Damianidi L, Kourlaba G (2013) A systematic review of the effectiveness of taxes on nonalcoholic beverages and high-in-fat foods as a means to prevent obesity trends. Clinicoecon Outcomes Res 5: 519-543
McCarthy M (2016) US food subsidies fuel obesity. BMJ 354: i3717
McLaren L (2007) Socioeconomic status and obesity. Epidemiol Rev 29(1): 29-48
Müller-Riemenschneider F, Reinhold T, Berghofer A, Willich SN (2008) Health economic burden of obesity in Europe. Eur J Epidemiol 23(8): 499-509
Nath T (2019) The economic cost of an obese society. https://www.investopedia.com/articles/personal-finance/041715/economic-cost-obese-society.asp; accessed on 18th September 2019
Roehling MV (1999) Weight-based discrimination in employment: psychological and legal aspects. Personnel Psychol 52: 969-1017
Schulte P. A., Wagner G. R., Ostry A., Blanciforti L. A., Cutlip R. G., Krajnak K. M., et al. (2007). Work, obesity, and occupational safety and health. Am. J. Public Health 97 428–436.
Sharkey JR, Johnson CM, Dean WR (2011) Less-healthy eating behaviors have a greater association with a high level of sugar-sweetened beverage consumption among rural adults than among urban adults Food Nutr Res 55: 10.3402/fnr.v55i0.5819
U.S. Centers for Disease Control and Prevention, Prevalence of Obesity in the United States, 2009-2010 (Atlanta: U.S. Centers for Disease Control and Prevention, 2012). www.cdc.gov/nchs/data/databriefs/db82.pdf; accessed on 27th September 2018

Obesity and Skin Problems

3

Prasanna Kannan Kutty and Methil Kannan Kutty

Abstract

Worldwide, obesity is a major health problem. It contributes to physiological modifications and the development of chronic diseases such as diabetes mellitus type 2, hypertension, cancer, etc. Obesity cases are only increasing rapidly in Western countries like the United States and Europe but also in Asia and Middle-East countries. An unhealthy diet, lack of physical activity and genetic defects are causes of obesity. Both genders are affected by obesity. People who are aged between 40 and 60 years old suffer from obesity. For some inexplicable reasons, the subject of obesity and its link with dermatological disorders has not attracted as much attention as it justifiably merits. However, myriads of studies indicate that obesity is certainly related to numerous skin problems such as *Acanthosis Nigricans*, *Striae distensae*, pressure ulcer, diabetic foot ulcer, lymphedema, Massive Localized Lymphadema (MLL), *Candida albicans* infections, Psoriasis (PSO), Erythrasma, Granular parakeratosis (GP), Pretibial mucinosis and Dercum disease. In this chapter, obesity-related skin diseases are discussed briefly.

Keywords

Obesity · Skin infections · Skin problems · Diseases

P. K. Kutty
Pantai Hospital, Kuala Lumpur, Malaysia

M. K. Kutty (✉)
Department of Pathology, Faculty of Medicine, Lincoln University College, Petaling Jaya, Selangor, Malaysia

M. K. Kutty, A. Elengoe (eds.), *Obesity and its Impact on Health*,
https://doi.org/10.1007/978-981-33-6408-0_3

3.1 Introduction

Obesity is a rapidly escalating global menacing malady and pandemic. It causes physiological modifications and is responsible for the development of gamut of diseases including obesity-linked comorbidities. For some inexplicable reasons, the subject of obesity and its link with dermatological disorders has not attracted as much attention as it justifiably merits. However, studies have revealed that obesity is associated with numerous skin problems. Obesity impacts many of the skin functions such as the skin barrier function, functions of the sebaceous glands and sebum production that preserves the skin moisturized and supple, sweat glands, subcutaneous small lymphatics, collagen configuration and function, besides others (Obesity and skin problems 2018). The epidermis, enriched with numerous proteins and other molecules support many defensive functions. There are many other noteworthy functions which include, thermal, immune, antioxidant, antimicrobial and penetrability blockade (Albanesi et al. 2005).

The keratinocytes are synthesizing and secreting many known inflammatory mediators; also regulate the skin's immune reactions (Lee et al. 2006). Aside from these keratinocytes express UV-absorbing molecules. The keratinocytes have vital role functions in thermal and UV-barrier functions (Hearing 2005). The cutaneous antimicrobial systems are principally intermediated via the surface lipids, skin surface acidification, iron-binding proteins and antimicrobial peptides (Bowdish et al. 2005; Braff et al. 2005). The antimicrobial peptides are once considered as an evolutionarily antique system of immune defense with barely any pertinence to the normal function of human skin. Current advances have highlighted that human skin expresses these gene-encoded peptides. The presence of these cutaneous peptides constitutes a barricade for inherent host safeguard against pathogenic microbes. Furthermore, antimicrobial peptides function by stimulating them to transform chemotaxis and chloride secretion (Gallo and Huttner 1998).

A brief supplementary elaboration of the functions of the skin is pertinent in this context. The permeability barrier function of the skin constitutes the most essential protective function. The multi-layered stratum corneum (SC) consisting of flattened, enucleate corneocytes, enclosed by multiple planar lamellae sheets, enclosing ceramides, cholesterol, and free fatty acids (FFA) (Elias and Menon 1991). The sites of these extremely hydrophobic lipids within the stratum corneum SC restrain the external migration of water (Elias and Menon 1991). The lipid organization determines the skin barrier function. By virtue of the remarkable stratum corneum lipid composition, the lipids behave differently from that of other biological membranes. In addition to a predominant SC crystalline phase, there is a lipid phase probably consisting of a subdivision of lipids. Both the crystalline and the lamellar phases are recognized as critical for the skin blockade function (Bouwstra et al. 2003).

3.2 Role of Epidermis

Besides the penetrability obstructing role, the SC has additional crucial roles as antimicrobial protection, hydration of viable epidermis, antioxidant defense and the advancement of a mechanical barrier. Because of The constitutional arrangement either corneocytes or the SC intercellular lipids arbitrate the defensive roles of the SC. Current studies have shown that these roles are so systematically and uniformly linked that any functional alteration modifies other functions. These defensive roles are jointly connected to many mechanisms (Albanesi et al. 2005).

The lamellar body (LB), in the epidermal keratinocyte, are viewed as essential in epidermal passable blockade as well as exfoliation. It is established that skin surface pH, ranging between pH 4.5 to pH 5.5 is vital for the epidermal permeability barrier, function as well as the epidermal antimicrobial blockade and SC integrity and structure (Parra and Paye 2003).

3.3 Sebum, Skin and Hair Texture

Sebaceous glands are ubiquitous in the distribution in the human body save the palms and soles. The precise function of these glands is still uncertain but many theories have been postulated such as antioxidant and antibacterial properties, and transportation of pheromones. The lipids generated by Sebaceous glands are implicated in the pathogenesis of acne in adolescents. While the bulk of lipids are of sebaceous gland origin they are also produced elsewhere in the body, wax esters and squalene are unique to the sebaceous gland (Smith and Thiboutot 2008).

Variations in sebum secretion between individuals are accountable for diverse skin quality. Other than its role as a protective barrier that lessens skin dryness sebum also has antibacterial properties, by virtue of the action of oleic and palmitoleic acids, and anti-oxidant properties. The development of acne is the result of blockage and subsequent infection of the sebaceous channels. Obesity-linked disorders decidedly aggravate the development of acne. Obese individuals have elevated androgens, insulin, growth hormone, and insulin-like growth factors (Lu et al. 2017).

3.4 Obesity and Lymphatics

Earlier studies-both clinical and laboratory have revealed that obesity negatively regulates lymphatic function (Ogden et al. 2014; Savetsky et al. 2014). Yet, it awaits confirmation whether the perceived impact on lymphatic function is attributable to weight gain, protracted exposure to high-fat diet (HFD), or toxic metabolic by-products of adipose tissue. The study by García-Toro et al. (2016) proposes that obesity-provoked tissue alterations or modifications in glucose and/or FA metabolism (or a blend of all) are responsible for lymphatic dysfunction but sustained administration of HFD per se does not directly damage the lymphatics.

HFD provoked obesity is found to categorically lessen interstitial fluid transport, immune cell swapping and lymphatic collecting vessel propelling (Weitman et al. 2013). These hostile sequelae imply that surplus adipose accretion participates in this reaction Unlike HFD, high fructose also diet-induced metabolic syndrome (Zawieja et al. 2011). Excessive blood cholesterol causes lymphatic system abnormalities which are autonomous of obesity.

3.5 Collagen in Skin

Scant information is available with regards to biomechanical studies on the skin of patients with massive weight loss (Choo et al. 2010) found no variation in tensile strength between abdominal skin of patients with considerable weight loss and those with normal weight. In fact, the medial abdominal skin of patients with massive weight loss was more robust than control skin from the same region in both vertical and horizontal directions. The skin of patients with massive weight loss was thicker than that of the normal weight group with an augmented thickness of both epidermis and dermis with thicker collagen fibers (Choo et al. 2010). Morbid obesity exhibited little impact on the skin quality because it looked nearly normal with good collagen fiber thickness and density, and only the elastic fiber network was weakened most likely as a result of undue stretching applied to the skin. It was obvious that the skin of patients with considerable weight loss is much feebler and less resistant, and it is deficient in the integrity of epidermis and dermis with extremely destroyed collagen and elastic fiber networks (Sami et al. 2015). The remarkable distinctive biochemical properties of the skin permit it to guard and adapt to the body that it covers. Essentials constituting the dermis-collagen and elastin-principally confers these properties. A clear comprehension of the biomechanical properties of skin is vital to achieve ideal cosmetic results in dermatologic surgery. The following are some of the dermatological entities associated with obesity.

3.5.1 Skin Tags (STs)

While skin tags are regarded as the most frequent fibrous lesions of the skin, they have not attracted as much attention it deserves in the dermatological literature; and what is more disconcerting is that hitherto, their precise etiology remains unclear (El Safoury et al. 2010). The literature contains reports of its link to obesity, diabetes mellitus, friction, acromegaly, organ transplant, human papillomavirus and other conditions.

Skin tags (STs) also known as soft fibromas, fibroepithelial polyps, or acrochordons denote a common benign dermatological disorder that presents as small pedunculated growths from the contiguous skin (Millington and Graham-Brown 2010). Microscopically STs appear as a polyp composed of loose, edematous fibrovascular core enclosed by slightly abnormal thickened epidermis; the center displays mild chronic inflammatory infiltration and dermis shorn of nerves. They frequently occur in—creases and folds of skin exposed to continuous abrasion (Cutaneous tag skin 2018). The incidence of ST has been appraised to be 46% in the general population (Acrochordon 2018). Their incidence is generally higher in women than in men. It is a matter of interest and consequence that *Molluscum contagiosum*, also of a viral etiology occurs in comparable areas as those of skin tags and grossly resembles skin tags. Interestingly, obese subjects' exhibit diverse colored STs' (El Safoury and Ibrahim 2011).

3.5.2 Pathogenesis of STs

Studies have revealed that low-risk HPV viruses are involved in the etiopathogenesis of skin tags (Gupta et al. 2008). Their occurrence usually exceeds in women compared to men. It exhibits a rare link with certain conditions. It is believed that a genetic component is involved in the etiopathology of skin tags. STs may be emblematic of those individuals with disturbed carbohydrate metabolism and for those who may be prone to diabetes mellitus type 2, although such a link has not been proven. Hyperglycemia and insulin escalate the prevalence of STs through a mechanism as yet undetermined (Tamega et al. 2010). Obese subjects displayed multicolored STs (El Safoury and Ibrahim 2011). An increase in the number of STs' was found to be linked more to obesity than DM, a finding in keeping with the results of Bhargava and Mathur (1996). The latter reported increases in the number of acrochordons matched with the severity of obesity and were consistent with the observations of Sari and Balci (2006). On the contrary, this study is not concurrent with the findings of Rasi et al. (2007).

3.6 Acanthosis Nigricans (AN)

Acanthosis nigricans (AN) was originally documented over a century. Paul Gerson Unna from Hamburg and reported a case in 1891. Vitezslav Janovsky from Prague followed suit reporting an analogous case (Schwartz 1994). In 2000, the American Diabetes Association had documented AN as a hazard for the development of diabetes in children (Sinha and Schwartz 2007). AN manifests as velvety, hyperpigmented plaques on the skin. Intertriginous areas are commonly affected. Other skin sites or mucosal surfaces may also affect. AN can occur in relation to diverse systemic disorders of which; obesity and diabetes mellitus are the most frequent. AN is associated with epidermal keratinocyte and dermal fibroblast proliferation. Of many types, the benign form is possibly connected with factors that provoke the epidermal cell proliferation. Other prospective mediators include epidermal growth factor receptor (EGFR) or fibroblast growth factor receptor (FGFR). Insulin at elevated concentrations possibly impacts through high-attraction fastening to IGF-1 receptors. Besides, obese patients with hyperinsulinemia exhibit elevated levels of free IGF-1 levels leading to enhanced cell development (Higgins et al. 2008).

3.6.1 Epidemiology of Acanthosis Nigricans

AN shows no gender or age predilection; it affects infants, children, and adults. Although the incidence of AN is reported variously, it is obvious that it occurs in a considerable percentage of obese and diabetic subjects; its occurrence may vary (Hud et al. 1992). AN seems high among people of Native American, African-American and Hispanic lineage compared to others.

3.6.2 Pathogenesis of Acanthosis Nigricans (AN)

The pathogenesis of *acanthosis nigricans* is still not well defined and is probably linked with interplay of factors as well as insulin-mediated activation of IGF receptors on keratinocytes, and elevated growth factor levels. Augmented circulating insulin stimulates keratinocyte insulin-like growth factor (ILGF) receptors, especially IGF-1. High levels of circulating IGF may cause keratinocyte and dermal fibroblast proliferation. Excessive sweating or rubbing may predispose to AN Risk factors for AN include; (a) obesity (b) insulin resistance (c) positive family history of AN or genetic syndrome (d) family history of diabetes mellitus and (e) malignancy (Higgins et al. 2008).

There is a link between AN and metabolic syndrome (MetS) (Furukawa et al. 2017). MetS is typified by the occurrence of multiple risk factors of which is type 2 diabetes mellitus. In fact, AN may be one of the presenting complaints of MS.

AN is also linked to many drugs which include nicotinic acid, glucocorticoids, diethylstilbestrol, oral contraceptives, growth hormone, estrogen, niacin and insulin. The incidence of obesity-linked AN is affirmatively associated with niacin use (Li et al. 2011). Excessive sweating as a result of physical activity helps excrete xenobiotics reduces the risk of AN. It is postulated that xenobiotics intensify cell proliferation in AN and probably compensates for prolonged xenobiotic exposure, however, this awaits further corroboration of this hypothesis. Long-term exposure to xenobiotics intensifies the risk for both AN and MetS (Otani 2011). The remarkable upsurge in xenobiotic exposure in the general population is attributable to food additives and specially compulsory performance of synthetic vitamin enrichment AN is also linked to many drugs which include nicotinic acid, glucocorticoids, diethylstilbestrol, oral contraceptives, growth hormone, estrogen, niacin and insulin (Li et al. 2011).

3.6.3 Fibroblast Growth Factor Receptor (FGFR)

Hereditary variants are related to fibroblast growth factor receptors (FGFR) defects. FGFR3 mutations reveal the potential for an extensive array of FGFR manifold effects besides locus heterogeneity in Crouzon syndrome. Crouzon syndrome is a distinctive disorder with widespread *acanthosis nigricans* (Mir et al. 2013).

3.6.4 Transforming Growth Factor (TGF)

Increased TGF seems to be the mechanism underlying malignancy-related AN. TGF acts on the epidermis through the FGFR. In this context it is pertinent to allude to AN and obesity syndrome, AN is a notable initial indicator of the obesity syndrome. AN heralds the risk of developing obesity syndrome. Recognition of AN consequently is important for preventive medicine (Maitra and Rowland 2004).

3.7 Striae

This strange disorder is believed to be due to the distending effect of the skin, with an associated collagen loss. *Striae distensae* are an extremely frequent form of dermal scarring that manifests on the skin as erythematous, violaceous, or hypopigmented linear striations. It is also known as striae, stretch marks, and *striae atrophicans. Striae distensae* due to pregnancy-related abdominal distension is aptly called *Striae gravidarum* (Elsaie et al. 2009). Two main forms of *Striae distensae* (SD) exist (a) *Striae rubra* and (b) *Striae alba. Striae rubra* are the initial manifestation of *Striae distensae*. They are symbolized by an erythematous to the violaceous colored lesion. In course of time, *Striae rubra* progress into *Striae alba*, which present as atrophic scar-like hypopigmented lesions. Common sites of its occurrence include the abdomen, breasts, medial upper arms, hips, lower back, buttocks and thighs. Although characteristically asymptomatic, SD besides defacing causes a severe impact on the psychology of patients. *Striae distensae* (SD) affect the skin as a result of constant and progressive distending with augmented stress on the connective tissue subsequent to enlargement of the different parts of the body. The sites of its occurrence differ depending on the underlying conditions; in pregnancy the abdomen and the breasts, in bodybuilders the shoulders, in adolescents going through their growth spurt and in overweight individuals. Although purple or red abdominal striae are an outstanding physical sign of Cushing's syndrome, they are not distinctively pathognomonic of this disease as they also occur infrequently in pregnant women, obese normal individuals and adolescent children. The latter three conditions are associated frequently with white or colorless striae.

3.7.1 Epidemiology and Risk Factors of Striae

Epidemiologic statistics are startlingly few and far between on SD in the overall population (Gilmore et al. 2012). As per available reports, frequencies have ranged from 11% in normal men up to 88% in pregnant females (Osman et al. 2007). Predisposing factors comprise pregnancy, adolescent age, drug exposure (e.g. topical or systemic corticosteroids), underlying disease (e.g. Cushing's syndrome) and cosmetic surgery (e.g. breast augmentation) (Basile et al. 2012).

3.7.2 Pathogenesis of Striae

While the precise factors in the pathogenesis of striae are yet to be fully understood, the SD represents the fragmentation of the connective tissue. Skin stretching may result in undue mast cell degranulation with consequent injury to collagen and elastin. Continued use of oral, topical corticosteroids or augmented adrenal cortical activity as in Cushing syndrome is implicated in the pathogenesis of the development of striae. Although participation of genetic factors is a distinct possibility, it awaits further understanding. This is further substantiated by the genome-wide association studies which provided proof of involvement of elastic microfibrils in the pathogenesis of non-syndromic *striae distensae* (Tung et al. 2013).

3.8 Pressure Ulcers and Obesity

While the definition of pressure ulcer is well known to all, still it bears repeating for the sake of academic satisfaction.

A pressure ulcer is defined as a local discontinuity of skin commonly due either to stress compounded by cut (Hyun et al. 2014). Ostensibly body mass index (BMI) is linked with such ulcers in intensive care units. Skinny or exceptionally fat patients have experience frequency of pressure ulcers relative to patients with normal weight or obese groups. This finding is in keeping with an incidence survey (Drake et al. 2010). While excessive obesity is correlated with its prevalence seem to safeguard patients from pressure ulcers. This underscores the need for considering the range of BMI value in evaluating the hazard for pressure ulcers. It is noteworthy that morbidly obese patients were 4–5 years younger and more likely to occur in women (Hyun et al. 2014).

While we are dealing with pressure ulcers, it is pertinent to allude to the development of other types of ulcers associated with and caused by obesity.

3.9 Diabetic Foot Ulcers

It is too well known to repeat that obesity is an important hazard for type II diabetes (Lowe 2009). The usual history of foot ulcer in diabetic patients stems from the impact of patient's contact with the surroundings The risk factors for foot ulcer include pathologic changes involving nerves, or vessels and foot injury (Misliza 2009). The incidence of diabetic foot ulcers is estimated to be about 15% of diabetic patients and this is a matter of great anxiety. One of the main reasons for hospitalization for diabetic patient is foot ulcers and infection. The occurrence of bone infection, removal of affected limb, or fatality can be dismal problems of a diabetic foot ulcer (Saffle et al. 2008). It is observed that diabetes mellitus is a robust predictor of diabetic foot ulcers and patients with type II DM are almost three fold prone compared to those with type I DM and other studies also confirm this (Saad et al. 2013). It is possibly attributable to changes referred to already. The patient may be beset with be the lack of factors that help healing culminating in foot ulcers. Diabetic patients with neuropathy have a robust link with foot ulcers (Misliza 2009). Analogous findings have been reported by studies in Tanzania, Jordan and Egypt (Bakri et al. 2012). Severely diabetic patients are prone to develop microvascular complication and neuropathy, which may intensify the risk for development of foot ulcer because of heightened pressure load and shearing forces.

3.10 Venous Insufficiency Ulcers

Chronic venous insufficiency (CVI) of the legs frequently occurs in excessively fat people. There is a significant association between obesity and clinical intensity of chronic venous disease (Danielsson et al. 2002), who opined that overweight seems to be an autonomous hazard for skin changes. Morbidly obese patients are oddly predisposed to refractory ulcers. The most important risk factors for emergent chronic venous disease comprise old age, obesity, and affirmative family history. While chronic venous disease affects relatively more often in women, some studies do not agree with this. Pregnancy appreciably escalates a woman's risk of developing chronic venous disease. Patients with excessive body weight exhibit limb symptoms, typical of CVI, but nearly 2/3 of the patients adduced proof emblematic of this condition. Because of the correlation with mounting bodyweight, it signifies that it per se is responsible for the morbidity. The cause of obesity-related chronic venous insufficiency has been linked to an augmented intraabdominal pressure as a result of abdominal fat masses which engenders venous hypertension (Scholl et al. 2017).

3.10.1 Lymphedema

Lymphedema is caused by the dilation of lymphatics as a result of diminished lymphatic drainage. Lymphatic fluid accumulation causes distension lessened tissue perfusion and seepage of protein-rich lymphatic fluid into neighboring tissue. Bacterial infections are common and permanent tissue changes are manifested by hyperkeratosis (skin thickening) (Lowe 2009).

At this junction, it is relevant to refer to an interesting entity-skin disorder associated with lower limb edema and distichiasis. It is characterized by aberrant eyelashes. A family history of this condition with lower limb edema can find in older children or at puberty. This is associated with heterozygous FOXC2 which is the diagnostic feature.

Another interesting entity that is familiar to many is lymphedema is associated with post-surgical mastectomy.

3.10.2 Massive Localized Lymphadema (MLL)

MLL is commonly regarded as an uncommon condition associated with the soaring occurrence of morbid obesity, with persons categorized as exceptionally obese (BMI $\geq$40 kg/m^2), it appears the incidence of MLL also is mounting pari passu obesity and hence MLL may no longer be categorized as rare (Fife 2014). MLL symbolizes a non-malignant excessive growth of fibrous and swollen fibro-fatty tissue. It presents as a huge mass commonly occurring in excessively fatty persons. Its common site of occurrence is the inner aspect of the thigh but it can also occur in other parts of the body. The massive size, MLL is often confused with highly malignant soft tumour otherwise called pseudosarcoma. Although it can be diagnosed clinically, biopsy tissue examination will be helpful as confirmatory to represent MLL. Albeit not rare among the severely obese, MLL is still generally heightened lly misdiagnosed (Fife 2014).

While we are on the subject of lymphedema, a brief allusion to this rare entity obesity-linked abdominal elephantiasis is pertinent and justifiable. Although rare awareness of such conditions, needless to emphasize would go long way in enabling clinicians and dermatologists to arrive at correct diagnose and avert underdiagnoses or misdiagnoses. Abdominal elephantiasis is rare, but a disfiguring, distressing and progressive dermatological disorder. Chronic lymphedema coupled with repeated streptococcal or *Staphylococcus* infections are the cause of abdominal wall elephantiasis. With the progress of the condition, the patients may become incapacitated not to mention prone to increased risk of mortality. Ritesh et al. (2013) have reported three morbidly obese patients with this condition. The condition is associated with lymphatic edema of legs with cobblestone-appearance of the skin. Two patients exhibited localized manifestations. One patient had pus from the abdominal wall. All the patients on antibiotics treatment displayed a little initial favorable response and hence refused surgery.

3.10.3 Pathogenesis of MLL

Excessive adipose tissue in morbidly obese patients can cause lymphatic blockage. Due to obesity, localized reduction of blood flow to the affected tissue and conceivably causes edema. Reduced blood flow may instigate a gush of changes, coupled with the enrolment of growth factors, simulating the repair process and culminating in the deposition of conspicuous fibrous tissue and fatty tissue. The possibility of a malignant transformation of this condition is to be reckoned with. Ever since the original clinical recognition MLL, its clinical manifestations have turned out to be more diverse. Half of the patients in the experience of Carolein had unusual anatomical sites of its co-occurrence. Chopra et al. (2015) had reported that almost half the lesions of MLL occurred in uncommon sites. This is in keeping with the findings of others implying these appearances are no more infrequent. This is ever more significant for the awareness of clinicians to be able to correctly diagnose this condition (Fife 2014). Chopra et al. (2015) from his earlier review and contemporary study observed the increase in the number of cases reported from 41 to 65. MLL can cause health problems including the likely development of malignancy and a related 9% mortality. Nevertheless, patients are primarily obsessed with social and quality-of-life trepidations, besides limitation in mobility, social stigma, the problem of obtaining proper apparel, and tests of personal hygiene (Fife 2014). MLL is commonly regarded as an uncommon condition. With the soaring occurrence of morbid obesity, it appears to be anticipated the incidence of MLL also is mounting pari passu and hence, MLL may no longer be categorized as rare (Fife 2014).

3.11 Skin Infections and Obesity

Obesity has a robust albeit not yet exactly defined impact on the immune response via a diversity of immune mediators, which predispose to infections (Falagas and Kompoti 2006). A retrospective cohort study of 2-year period among the adult Saudi population revealed that the most frequently reported infections in obese men were tinea pedis (18%), intertrigo (14.7%), and bacterial folliculitis (13.1%) with significant difference (Kahlil et al. 2011).

Prior to the discussion of cutaneous candidiasis in obesity, it is relevant to briefly allude to the background of candida infection in general.

3.11.1 Candida Albicans

Candida albicans is a causative organism of a variety of distinct types of diseases. *Candida albicans* is a dimorphic fungus causing many diseases in humans. It is a normal commensal organism found in the gastrointestinal and reproductive mucosa, but in the immunocompromised state, *C. albicans* infections cause oral and oropharyngeal (OPC), vulvovaginal (VVC), mucocutaneous or disseminated candidiasis. It is notable that the immune reactions to these different disease states are to a certain extent dissimilar, indicating the intricacy of the anatomical basis for host resistance against fungus. Chronic mucocutaneous candidiasis (CMC) represents a gamut of chronic or recurrent diseases affecting the skin, nails (onchomycosis) and mucous membranes (Conti and Gaffen 2010). In many cases, CMC is secondary to autoimmune endocrinopathies, although this link is not fully understood (Eyerich et al. 2008). The transition from healthy to pathogenic state takes place at the mucosal-fungal interface. The last few years have witnessed a paradigm shift in our comprehension of the adaptive and innate immune mechanisms that bestows protection from candidiasis. Conventionally, defensive adaptive reactions counter to Candida were attributed to the mediation by CD4 Th1-effector cells (Dongari-Bagtzoglou and Fidel 2005). In 2005, a major revolution in the perception of T helper cell biology took place with the unearthing of a novel subset distinct from orthodox Th1/Th2 populations, which focused on the mutual IL-12p40 subunit. Regarded as "Th17 cells" because of the production of IL-17 these cells are produced subsequent to exposure to IL-1, IL-6 and TGF-β.

3.11.2 Cutaneous Candidiasis

Obese subjects who have deeper skin folds and excess sweating, which can give rise to friction, moisture and skin maceration; all of which provide a conducive cutaneous melieu susceptible to infections from yeast, bacteria and fungus. The intertriginous regions which are habitually moist favor the growth of yeasts, such as Candida spp., it manifests as typical satellite lesions with erythema and irritation in the inframammary region, beneath the abdominal fold(s) and in the inguinal area. *Candida folliculitis*, which is less frequent, poses a problem to obese patients. Patients are also prone to have tinea cruris (ringworm) infection and intertrigo (Hoste et al. 2013).

Recent developments have revealed that in bacterial-fungal cross-talk, bacteria can restrict or support the fungal population and vice versa (Morales and Hogan 2010). Numerous studies have described interactions between *C. albicans* and its bacterial neighbors (Morales and Hogan 2010). Concurrent bacterial and fungal infections have been implicated in intensified host colonization and virulence. This is exemplified by the supportive interaction between *C. albicans* and *Escherichia coli* which augments adhesion of *C. albicans* to bladder mucosa increasing the possibility of fungal urinary tract infections. Similarly, the hazard of *Pseudomonas aeruginosa* induced pneumonia is strikingly more in patients harbouring *C. albicans* (Azoulay et al. 2006). Microbial pathogens need to integrate vital micronutrients to launch infections. Microbes including fungi have to control their intracellular levels of transition metals such as iron, copper, zinc to propitiate their physiological requirements while evading harm. Hence, expectedly the host exploits both the usefulness as well as the harmfulness of transition metals to protect against microbial incursion. There is homeostasis at the pathogen-host interface. This symbolizes a part of nutritional immunity (Hood and Skaar 2012).

C. albicans has a repertoire of factors that contribute to survival, persistence and virulence; among this array of capability and virulence characteristics are iron procurement factors and pathways which permit fungal cells to obtain this essential mineral in the host's iron-poor milieu. Because of the link of iron in the human host to host proteins, pathogens such as *C. albicans* must possess mechanisms to gain iron from these proteins. The potential iron sources for invading microbes are the important human proteins, including haemoglobin, transferrin, lactoferrin and ferritin, which contain iron (Almeida et al. 2009). The fungus reacts by modifying its iron absorption to the declined iron levels. It is puzzling that iron levels rise in other parts of the kidney with the progress of Candida infection. Iron is an indispensable nutrient for *C. albicans*, and iron uptake may play a special role in supporting virulent infections. The bulk of iron in a healthy human exists as heme in hemoglobin and uptake systems for heme iron for *C. albicans* possess a hemolytic activity, and plasma membrane proteins are able to bind hemin and hemoglobin (Pendrak et al. 2004). The uptake pathway for hemin iron relies on the intracellular heme oxygenase (Hmx1p) to free iron from the porphyrin chelate. Candida infection disrupts overall iron homeostasis in the host by interrupting splenic erythrocyte recycling and this is linked with augmented iron storage in the kidney medulla.

Consequently, fungal infection exerts a system-wide impact on iron homeostasis in the mammalian host whilst activating local nutrient immunity to restrict the infection (Potrykus et al. 2013).

Considerable progress in the insightful comprehension of immunity associated with fungal pathogens has been achieved remarkably in recent years. Especially, notable is the comparable findings in the C-type lectin receptor family and the Th effector arms of immunity, especially Th17 cells and their marker cytokine, IL-17. Many of these studies have focused on the most common human fungal pathogen, *C. albicans*, which is a classical commensal microbe in healthy people but is causally responsible for several diseases in immunocompromised hosts, ranging from minor mucosal infections to deadly pervasive disease (Hernández-Santos and Gaffen 2012). Th17 cells are induced with transforming growth factor (TGF)-beta and IL-21, IL-6 or IL-1β and play a vital role in combating extracellular pathogens (Kleinewietfeld and Hafler 2013). Th17 cells were initially described based on their secretion of IL-17. They express a series of other cytokines including IL-17F, IL-21, GM-CSF and IL-22. Their ancestral specific transcription factor which is the retinoic acid receptor-related orphan receptor γt (RORγt, in humans RORc); controls development and function of Th17 cells.

3.12 Psoriasis (PSO)

Psoriasis (PSO) is a chronic relentless often relapsing skin disease. This disease affects about 2–3% of the global population, although it is more prevalent in the native population of the United States, Canada, and Europe (Kim and Krueger 2015). In the beginning PSO was regarded exclusively caused by malfunctioning of strikingly multiplying keratinocytes. It was perceived that presence of immune cells infiltration which does not constitute a major factor in pathogenesis is only the result of hyper proliferating. Nevertheless, the positive amelioration of PSO following immunosuppressive therapy, for instance, Cyclosporine, denileukin, diftitox and alifacept endorsed the immune synchronization in the pathogenesis of PSO (Colombo et al. 2013). In due course, the impact of various immune factors was better clarified. The finding of abundant T cells especially Th1 and Th17 cells in PSO lesions was convincing the significance of the role of immune mechanisms (Lowes et al. 2014). Moreover, TNFα, iNOS generating inflammatory dendritic cells (DCs) densely infiltrate psoriatic skin; these DCs can separate Th1 and Th17 cells (Colombo et al. 2013). Finally, psoriatic skin is permeated by countless other immune cells. In addition, increased endothelium lining cells is visible. These cells possibly may have a contributory role in PSO pathogenesis. It is of interest to note that the first identified gene PSORS 1, was notably linked with PSO vulnerability was HLA-Cw6 sited at 6p.3 (Lowes et al. 2014). Obesity not only renders people more vulnerable to develop PSO but also intensifies PSO. Obesity must be an autonomous risk factor for PSO (Ferreli et al. 2018). Psoriasis causes considerable bodily and emotional agony and harms generally leading to an adverse effect on patient quality of life.

Conceivably major ominous effect of psoriasis is the concomitant comorbid conditions of obesity, for instance, diabetes, cardiovascular disease, which lessen lifespan. Yet another grim complication is psoriatic arthritis.

3.12.1 Pathogenesis of PSO

This subject is very extensive and we have only touched on certain salient aspects (Ferreli et al. 2018). PSO is conceivably the result of improper immune reaction to causal agents such as self-antigens, or non-self-antigens. It is the enrolment of leucocytes and the instigation of the lesions that continue the conservation of PSO. Earlier studies have revealed the role of chemokines in psoriatic inflammation and the significance of chemokine ligands and receptors in leukocyte enrolment of T cells into the psoriatic lesion. There has been a significant insight into the realm of certain chemokines in PSO as CX3CLI and CX3CR1. It has been hypothesised that predilection for the occurrence of PSO. Single nucleotide polymorphisms (SNPs) are linked with cytokine gene polymorphisms. It has been noted that SNPs within CX3CR1are linked with PSO. PSO seems to belong to certain autoimmune diseases. PSO triggers the innate immune system with which most of the vulnerability loci. It is now acknowledged that the disease arises from disruption of immune tolerance (Fry et al. 2015).

For decades there has been much dispute on the issue of pathogenesis of Psoriasis (PSO). The debate was centered on its basic nature and its relation to the faulty immune scheme. The questions that were rose 'Is it of the infective or microbial origin or autoimmune basis?' (Nickoloff and Nestle 2004). It has been proposed that a spectrum of cells coupled with a range of Th1 type cytokines and definite growth factors have major participation in the pathogenesis of PSO. The cells referred to a host of cells including T lymphocytes, monocytes, B cells, keratinocytes and natural killer cells in addition to many others.

Briefly, allude to the role of T lymphocytes in the pathogenesis of PSO. It is generally supposed that the pathogenesis of PSO involves T cell anomaly in conjunction with collaboration between keratinocytes and intricate cytokine system. Lest the major fault is that of keratinocytes, any form of damage to be it physical or otherwise to the malfunctioning keratinocytes may trigger the creation and liberate cytokines thus leading to antigen-free instigation of T lymphocytes. This would cause the additional discharge of cytokines with subsequent propagation of both keratinocytes and T lymphocytes; furthermore, it will lead to inflammation. Contrasting normal skin, cytokines liberated by psoriatic skin enhance triggering of T lymphocytes which stimulate increased keratinocyte multiplication by virtue of their explicit receptors. Some studies have incriminated alterations of the basal lamina and a multifaceted complex of cytokines, chiefly Th1 type, are implicated in various phases of pathogenesis.

It is important to record that in human skin keratinocytes differentiate abnormally in the hyperproliferative epidermis. Psoriasis, a hyperproliferative dermatological disorder, is associated with amplified cell multiplication and the belated start of epidermal differentiation. Unlike the normal epidermis, the cell cycle takes place in 4 days in the psoriatic epidermis. Growth factors, originating from diverse cell types, are thought to control the augmented multiplication.

3.13 Cytokine Mediators and Their Significance

Even though a multifaceted system of numerous cytokines has been implicated in the pathobiology of psoriasis, none of this perse can be regarded to be contributory. While it is perceived that keratinocytes constitute a chief basis for pro-angiogenic cytokines (VEGF, IL-8) the exact underlying mechanism for neo vasculogenesis in psoriasis remains indistinct. Initial psoriatic plaque exhibits migration of stimulated endothelial cells which put down BM. The lesional vessels articulate E-selectin, facilitating the intra-cutaneous movement of leukocytes (Franziska et al. 2020).

Yet it must be acknowledged that angiogenesis may not end all and be all in the pathogenesis of psoriasis. Vitamin D analogues, retinoid, and cyclosporine have anti-vasculogenic, anti-propagatory and anti-inflammatory properties. Even though a multifaceted system of numerous cytokines has been implicated in the path biology of psoriasis, none of this perse can be regarded to be contributory.

3.14 Psoriasisi and Susceptibility Loci Comprising

3.14.1 Hidradentits Suppurativa (HS)

HS as the name implies, is an inflammatory cutaneous disorder with typical presentation. The presentation includes repeated or persistent suppuration in the region of sudoriferous glands. Hence, the common sites of occurrence include inframammary folds, axilla, groin, perineal or perigenital regions. Apparently, the disease shows no special prediction for age or gender. Because of its occurrence in groin, perineal or perigenital areas, it may be confused with lymphogranuloma venereum.

A spectrum of conditions (1) furuncles (follicular pyoderma) (2) carbuncles, (3) abscesses, (4) acne, (5) cutaneous manifestation of regional ileitis. (6) cutaneous tuberculosis, (7) actinomycoses and (8) steatocystoma multiplex can resemble HS. The details of the criteria of the differential diagnosis are not discussed because of space constraints. The readers will surely appreciate this regrettable omission.

Although the exact cause of HS remains undetermined, the past several studies, have therefore postulated that the disease is instigated by hereditary and ecological factors) (Zouboulis et al. 2015). Prominent environmental hazard factors in the pathogenesis of HS comprise cigarette smoking, obesity, and overweight. It is equally noteworthy the study of the role of genetic factors is underscored demonstrating that a sizeable proportion of HS patients have an ancestral history of HS. Ancestoral forms of HS exhibit an autosomal dominant heritage pattern (Scheinfeld 2013).

The basic flaw in HS encompasses obstruction with consequent folliculitis compounded by both innate and adaptive immune abnormality (Prens and Deckers 2015). Secondary bacterial infection occurs culminating in the formation of abscess which destroying first the pilosebaceous component and subsequently other adnexal structures. Additional contributory factors embrace changed manifestation of a spectrum ranging from antibacterial proteins' aberrant sudoriferous gland secretion, atypical epidermal inversions climaxing in sinuses, and absence of sebaceous glands (Kamp et al. 2011).

Patients reveal transformed innate and adaptive immunity, obligatory for microbial regulation (Wolk et al. 2017). Patients exhibited a comparative lack of intralesional anti-microbial proteins (AMPs) and an affirmative association between IL-22 and IL-20 in the lesions against AMP levels. On the other hand, Emilanov et al. (2012) identified using immunohistochemistry, a marked appearance of cathelicidine LL37 in the lesions. Besides, epidermis and dermis showed enhanced AMP human β-defensin 3. The AMP variations in HS remain unclarified. The role of clinical inflammation is unclear. Is it initiator or aggravator is contentious. Lack of AMP is an inconstant characteristic of HS and its evolution to progressive phases is uncertain (Wollina et al. 2013).

In particular, TNF-α, is a major participant, in the development of HS along with an affirmative association with its rigorousness (Mozeika et al. 2013). It is obvious that this condition exhibits flaws in keratin removal, with consequent follicular obstruction, and defective inbred cellular immune response. While Notch signaling is essential for modifying delineation of the epidermis and hair follicles, the exact role of this signaling in advanced phase epidermal demarcation. Deficiency of the notch signaling pathway transforms epidermal cysts and compromises sudoriferous gland equilibrium prompting inherent invulnerability, backing up sustaining persistent inflammatory reaction. According to Rangarajan et al. (2001), the vital role of TNF-α in the development of HS is illustrated by its overexpression seen in HS, coupled with showing an affirmative link with disease intensity (Mozeika et al. 2013), validating that HS as a follicular disorder involving a certain flaw in keratin clearance. It merits mention here; current proof has adduced that disturbed immune reactions participate in the pathogenesis of HS. It has been observed that TNF gene-related SNPs implies HS vulnerability for the response to TNF antagonists (Savva et al. 2013).

3.15 Erythrasma

A brief consideration of erythrasma is relevant in this context; especially because of its link with obesity. Erythrasma denotes an unremitting surface infection of the interdigital areas of the skin. Erythrasma manifests as well-demarcated red or brown patches with fine scaling and superficial fissures. The lesions may be slightly pruritic. It is caused by *Corynebacterium minutissimum*, which is generally a commensal Erythrasma, a frequent dermatological disease is caused by microorganism. Von Barensprung labelled it as erythrasma and the microorganism was called

Microsporum minutissimum. Erythrasma usually causes infection of the interdigital areas of the body and causes pruritus, scaling, and erythema. It is often symptomless.

A case of erythrasma due to *Corynebacterium afermentans* has been recorded (Dellion et al. 1996). One study, isolated mixed colonies of *Corynebacterium aurimucosum* with *Microbacterium oxydans* from the interdigital area of the left foot in an elderly woman indicating the possible involvement of other species of microorganisms in this condition (Yasuma et al. 2011). Age may escalate the prevalence of erythrasma. Its occurrence is universal and men show a greater predilection for this.

3.15.1 Incidence and Risk Factors

One study in the UK, reported a 19% incidence of erythrasama while 20% frequency was recorded in an Auckland skin clinic. In general, ostensibly natives of African American descent are extremely vulnerable to this. Ecological and host factors escalate the hazard of erythrasma. Predisposing factors include residing in damp environs, extreme perspiration, diabetes, meager sanitation, elderly persons, concomitant dermatological diseases and obesity.

Erythrasma typically affects normal and healthy adults while children are less prone to develop this. Individuals residing in crammed areas such as institutions are also more susceptible to erythrasma. *C. minutissimum* grows in the superficial keratinising layer of the epidermis. It engenders a coral-red luminescence on Wood's light examination because of coproporphyrin III. This is of diagnostic significance. Immuno incompetent individuals are predicted to have problems which include: infection of the endocardium, pustule formation, soft tissue inflammation and infection of the pelvis of the kidney.

3.16 Granular Parakeratosis (GP)

The original description of Granular parakeratosis (GP), in 1999 as a non-malignant dermatological disease which presents uncommonly itchy, dark brown to reddish blotches and plates in cutaneous folds. GP is a rare form of parakeratosis most often seen in the axilla, although other cutaneous folds are affected. Its distinctive histopathologic profile with typical features of dense *stratum corneum*, intense parakeratosis with intact keratohyalin granules, vascular propagation, and expansion will permit making a proper diagnosis. GP has been associated with the extravagant application of topical preparations, particularly cosmetics; and deodourisers however, it occurs in persons who are non-users of such agents. GP is also linked with occlusive melieu, excess perspiration and occasional local irritation. Although irritant contact reaction seems causal, mechanical irritation may also have a part to play in provoking these cutaneous alterations. Obesity renders people vulnerable to GP. Some studies have projected that GP has congenital links, and others have regarded it as a malady or constitute a reaction. The origin and development of this condition are doubtful. Weaver et al. (2007) suggested that a fundamental fault lines in the dispensation transformation of pro-filaggrin, which helps retain the keratin in the epidermis. In view of the alleged link between GP and excessive application of topical preparations, a milieu conducive, and profuse perspiration, it is proposed to signify allergic response. While certain GP patients have been free from using topical preparations its etiological link to such preparations remains obscure. Of note is the finding of lack of an essential protease involved in the suitable build-up of a wholly functional skin obstruction, which is more likely to develop GP may shed light on the development of GP. Regardless, the principal basis for GP is still undetermined.

3.17 Pretibial Mucinosis

Cutaneous mucinoses denote an assortment of entities symbolised by augmented quantities of acid mucin in the sub-epidermal zone (dermis). They can be widespread or limited to a small area and arise *de novo* or may be associated with systemic disorders. Obesity-associated lymphedematous mucinosis is a unique variation of mucinosis developing in obese patients sans any thyroid disorder. Hitherto perusal of English and other publications reveals only a small number of cases of this unusual entity (Ferreli et al. 2018). The rarity of this entity is obvious from the small number of reported publications in the literature (Milman et al. 2016).

Pretibial mucin deposition occurs in association with thyroid diseases (pretibial myxedema) or without any thyroid dysfunction (pretibial mucinosis). Obesity-associated pretibial mucinosis, has typical histological and morphological features. The pathogenesis of this rare condition remains undetermined. Apparently, obesity-induced lymphatic stasis, with local anoxia causes mucin accretion (Milman et al. 2016). This seems to be the outcome of the interstitial accumulation of plasma proteins attributable to faulty lymphatic drainage and the reduction of blood flow supported by often related venous insufficiency. Obesity-associated lymphedematous mucinosis (OALM) is an infrequent and recently documented disease that can clinically simulate pretibial myxedema. The lack of thyroid disorder coupled with histopathological profile can distinguish between the two conditions. The rarity of the condition is strikingly obvious in view of a small number of cases in scientific publications in the literature. English literature reveals 24 cases of Ferreli et al. (2018) study has reported two cases—(1) a 76-year-old obese, diabetic woman with a prolonged history of hypertension and (2) a 72-year-old woman with heart OALM as the name implies symbolizes a form of pretibial mucinosis, sturdily associated with obesity. It has typical morphological features and histological features which typified by skin-coloured to reddish-yellow semi-transparent papules, flat plates and nodules, originating in a swollen skin with lymph stasis in the front of the tibia. The pathogenesis of this unusual condition is yet to be completely determined. Apparently obesity-induced lymphatic stasis, with local hypoxia causes mucin accretion (Milman et al. 2016). This seems to be the outcome of interstitial accretion of plasma proteins attributable to faulty lymphatic drainage and the ischemia supported often by related venous insufficiency. OALM, as the name implies, symbolizes a form of pretibial mucinosis, sturdily associated with obesity, Pretibial mucin deposition (pretibial myxoedema) occurs in association with thyroid diseases or exclusive of any thyroid dysfunction (pretibial mucinosis).

Obesity-related pretibial mucinosis and pretibial myxoedema have distinct histopathological features like venous insufficiency-associated dermal mucinosis. Hence, histological discrimination between the conditions will not present any problems. Obesity-associated pretibial mucinosis; (OALM) has classic morphological and histological features. Apparently, obesity-induced lymphatic stasis, with local anoxia causes mucin accretion (Milman et al. 2016). This seems to be the outcome of the interstitial deposition of plasma proteins attributable to faulty lymphatic drainage and reduced blood flow supported often by a related venous insufficiency. OALM and pretibial myxoedema have distinct histopathological features like venous insufficiency-associated dermal mucinosis.

3.18 Adiposis Doloraosa (Dercum Disease)

Adiposis dolorosa is an exceptional disease manifesting numerous distressing tender fatty nodules or masses beneath the skin. This disorder like a few other diseases has the distinction (regret the pun) of many synonyms encompassing Ander syndrome, morbus Dercum, adipose tissue rheumatism, adiposalgia, or lipomatosis dolorosa. It has been categorised into four distinct forms (Hao et al. 2018).

While the condition exact cause of the condition is indeterminate it has been presently postulated to include dysfunction of several systems in the body and neural and other injuries. It is regarded as an autosomal dominant disorder. The majority of cases occur randomly and have no genetic basis (Miraglia et al. 2013).

Patients who suffer pain are typically complained of as throbbing, aching, piercing, or scorching in quality and it is usually generally uniform. Classically, touching aggravates pain and pain symptoms are not commensurate with findings on examination. The pain is refractory to treatment. It is of special significance that patients are generally excessively fat. Most cases of the disease are found to be sporadic with no specific genetic mutations (Miraglia et al. 2013). Generally, the predicted sites of painful adipose tissue include the upper limbs, chest, and buttocks.

At this juncture, we are tempted to suggest to readers to fathom the significance of dismal pain in the monumental book nay memoir by the illustrious Professor Dr. Brand and Yancey (1997) who pioneered the most useful tendon transplant for Hansens disease patients. It is notable that the diagnosis of this condition is one of exclusion. This becomes very obvious from the wide spectrum of conditions that may resemble Dercum's disease.

3.19 Conclusion

This chapter has strived to analyse the impact of obesity on the skin. Many studies revealed that obesity has been associated with many skin problems. The author first started this chapter by looking at the skin such as the role of the epidermis; sebum, skin and hair texture; obesity and lymphatics; collagen in the skin; and then on the obesity-related skin problem diseases. The authors hope that this chapter has been able to shed some light on the latter as we all continue to play our role in tackling this pandemic.

References

Acrochordon (2018) https://emedicine.medscape.com/article/1060373-overview; accessed on 20th August 2018

Albanesi C, Scarponi C, Giustizieri ML, Girolomoni G (2005) Keratinocytes in inflammatory skin diseases. Current Drug Targets-Inflammation & Allergy 4(3): 329-334

Almeida RS, Wilson D, Hube B (2009) *Candida albicans* iron acquisition within the host. FEMS Yeast Research 9(7): 1000-1012

Azoulay E, Timsit J-F, Tafflet M, de Lassence A, Darmon M, Zahar J-R (2006) Candida colonization of the respiratory tract and subsequent pseudomonas ventilator-associated pneumonia. Chest 129(1): 110-117

Bakri FG, Allan AH, Khader YS, Younes NA, Ajlouni KM (2012) Prevalence of diabetic foot ulcer and its associated risk factors among diabetic patients in Jordan. Jordan Medical Journal 171 (785): 1-16

Basile FV, Basile AV, Basile AR (2012) Striae distensae after breast augmentation. Aesthetic Plastic Surgery 36(4): 894-900

Bhargava P, Mathur D (1996) Acrochordon, diabetes and associations. Indian Journal of Dermatology, Venereology, and Leprology 62(4): 226

Bouwstra JA, Honeywell-Nguyen PL, Gooris GS, Ponec M (2003) Structure of the skin barrier and its modulation by vesicular formulations. Progress in Lipid Research 42(1): 1-36

Bowdish DM, Davidson DJ, Hancock R (2005) A re-evaluation of the role of host defence peptides in mammalian immunity. Current Protein and Peptide Science 6(1): 35-51

Braff MH, Bardan A, Nizet V, Gallo RL (2005) Cutaneous defense mechanisms by antimicrobial peptides. Journal of Investigative Dermatology 125(1): 9-13

Brand P, Yancey P (1997) Gift of Pain, The Paperback. https://www.amazon.com/Gift-Pain-Paul-Brand/dp/0310221447; accessed on 1st September 1997

Choo S, Marti G, Nastai M, Mallalieu J, Shermak MA (2010) Biomechanical properties of skin in massive weight loss patients. Obesity Surgery 20(10): 1422-1428

Chopra K, Tadisina KK, Brewer M, Holton LH, Banda AK, Singh DP (2015) Massive localized lymphedema revisited: a quickly rising complication of the obesity epidemic. Annals of Plastic Surgery 74(1): 126-132

Colombo M, Cassano N, Bellia G, Vena G (2013) Cyclosporine regimens in plaque psoriasis: an overview with special emphasis on dose, duration, and old and new treatment approaches. The Scientific World Journal: 2013

Conti HR, Gaffen SL (2010) Host responses to *Candida albicans*: Th17 cells and mucosal candidiasis. Microbes and Infection 12(7): 518-527

Cutaneous skin tag (2018) https://medlineplus.gov/ency/article/000848.htm; accessed on 22nd December 2018

Danielsson G, Eklof B, Grandinetti A, Kistner R (2002) The influence of obesity on chronic venous disease. Vascular and Endovascular Surgery 36(4): 271-276

Dellion S, Morel P, Vignon-Pennamen D, Felten A (1996) Erythrasma owing to an unusual pathogen. Archives of Dermatology 132(6): 716-717

Dongari-Bagtzoglou A, Fidel Jr P (2005) The host cytokine responses and protective immunity in oropharyngeal candidiasis. Journal of Dental Research 84(11): 966-977

Drake DJ, Swanson M, Baker G, Pokorny M, Rose MA, Clark-Reed L (2010) The association of BMI and Braden total score on the occurrence of pressure ulcers. Journal of Wound Ostomy & Continence Nursing 37(4): 367-371

Elias PM, Menon GK (1991) Structural and lipid biochemical correlates of the epidermal permeability barrier. Advances in Lipid Research 24: 1-26

Elsaie ML, Baumann LS, Elsaaiee LT (2009) *Striae distensae* (stretch marks) and different modalities of therapy: an update. Dermatologic Surgery 35(4): 563-573

El Safoury OS, Ibrahim M (2011) A clinical evaluation of skin tags in relation to obesity, type 2 diabetis mellitus, age, and sex. Indian Journal of Dermatology; 56(4): 393

El Safoury O, Fawzi M, Hay RMA, Hassan AS, El Maadawi Z, Rashed L (2010) Increased tissue leptin hormone level and mast cell count in skin tags: A possible role of adipo immune in the growth of benign skin growths. Indian Journal of Dermatology, Venereology, and Leprology 76 (5): 538

Emelianov V, Bechara F, Gläser R, Langan E, Taungjaruwinai W, Schröder J (2012) Immunohistological pointers to a possible role for excessive cathelicidin (LL-37) expression by apocrine sweat glands in the pathogenesis of hidradenitis suppurativa/acne inversa. British Journal of Dermatology 166(5): 1023-1034

Eyerich K, Foerster S, Rombold S, Seidl H-P, Behrendt H, Hofmann H (2008) Patients with chronic mucocutaneous candidiasis exhibit reduced production of Th17-associated cytokines IL-17 and IL-22. Journal of Investigative Dermatology 128(11): 2640-2645

Falagas ME, Kompoti M (2006) Obesity and infection. The Lancet Infectious Diseases 6(7): 438-446

Ferreli CA, Pinna ALA, Pilloni LB, Corbeddu MA, Rongioletti FA (2018) Obesity-associated lymphedematous mucinosis: two further cases and review of the literature. Dermatopathology 5: 16-20

Fife C (2014) Massive localized lymphedema, a disease unique to the morbidly obese: a case study. Ostomy Wound Manage 60(1): 30-35

Franziska G, Andreas K, Edgar S, Matthias G, Khalid M (2020) Current developments in the immunology of psoriasis. Yale Journal of Biology Medicine 93(1): 197-200

Fry L, Baker BS, Powles AV, Engstrand L (2015) Psoriasis is not an autoimmune disease? Experimental Dermatology 24(4): 241-244

Furukawa S, Fujita T, Shimabukuro M, Iwaki M, Yamada Y, Nakajima Y (2017) Increased oxidative stress in obesity and its impact on metabolic syndrome. The Journal of Clinical Investigation 114(12): 1752-1761

Gallo RL, Huttner KM (1998) Antimicrobial peptides: an emerging concept in cutaneous biology. Journal of Investigative Dermatology 111(5): 739-743

García-Toro M, Vicens-Pons E, Gili M, Roca M, Serrano-Ripoll M, Vives M (2016) Obesity, metabolic syndrome and Mediterranean diet: Impact on depression outcome. Journal of Affective Disorders 194: 105-108

Gilmore SJ, Vaughan Jr BL, Madzvamuse A, Maini PK (2012) A mechanochemical model of *striae distensae*. Mathematical Biosciences 240(2): 141-147

Gupta S, Aggarwal R, Gupta S, Arora SK (2008) Human papillomavirus and skin tags: Is there any association? Indian Journal of Dermatology, Venereology, and Leprology 74(3): 222

Hao D, Olugbodi A, Udechukwu N, Donato AA (2018) Trauma-induced adiposis dolorosa (Dercum's disease). BMJ Case Reports: 2018

Hearing VJ (2005) Biogenesis of pigment granules: a sensitive way to regulate melanocyte function. Journal of Dermatological Science 37(1): 3-14

Hernández-Santos N, Gaffen SL (2012) Th17 cells in immunity to *Candida albicans*. Cell host & microbe 11(5): 425-435

Hoste E, Denecker G, Gilbert B, Van Nieuwerburgh F, Van Der Fits L, Asselbergh B (2013) Caspase-14-deficient mice are more prone to the development of parakeratosis. Journal of Investigative Dermatology 133(3): 742-750

Higgins SP, Freemark M, Prose NS (2008) *Acanthosis nigricans*: a practical approach to evaluation and management. Dermatology Online Journal 14(9)

Hood MI, Skaar EP (2012) Nutritional immunity: transition metals at the pathogen-host interface. Nature Reviews Microbiology 10(8): 525

Hud JA, Cohen JB, Wagner JM, Cruz PD (1992) Prevalence and significance of *acanthosis nigricans* in an adult obese population. Archives of Dermatology 128(7): 941-944

Hyun S, Li X, Vermillion B, Newton C, Fall M, Kaewprag P (2014) Body mass index and pressure ulcers: improved predictability of pressure ulcers in intensive care patients. American Journal of Critical Care 23(6): 494-501

Kamp S, Fiehn A, Stenderup K, Rosada C, Pakkenberg B, Kemp K (2011) Hidradenitis suppurativa: a disease of the absent sebaceous gland? Sebaceous gland number and volume are significantly reduced in uninvolved hair follicles from patients with hidradenitis suppurativa. British Journal of Dermatology 164(5): 1017-1022

Khalil GM, Al Shobaili HA, Alzolibani A, Al Robaee A (2011) Relationship between obesity and other risk factors and skin disease among adult Saudi population. The Journal of The Egyptian Public Health Association 86(3): 56-62

Kim J, Krueger JG (2015) The immunopathogenesis of psoriasis. Dermatologic Clinics 33(1): 13-23

Kleinewietfeld M, Hafler DA (2013) The plasticity of human Treg and Th17 cells and its role in autoimmunity. Seminars in immunology 2013: Elsevier

Lee SH, Jeong SK, Ahn SK (2006) An update of the defensive barrier function of skin. Yonsei Medical Journal 47(3): 293-306

Li X, Hu Z, Qu X, Zhu J, Li L, Ring BZ (2011) Putative EPHX1 enzyme activity is related with risk of lung and upper aerodigestive tract cancers: a comprehensive meta-analysis. PloS One 6(3): e14749

Lowe JR (2009) Skin integrity in critically ill obese patients. Critical Care Nursing Clinics 21(3): 311-322

Lowes MA, Suarez-Farinas M, Krueger JG (2014) Immunology of psoriasis. Annual Review of Immunology 32: 227-255

Lu L, Lai H, Pan Z, Wu Z, Chen W, Ju Q (2017) Obese/overweight and the risk of acne vulgaris in Chinese adolescents and young adults. Hong Kong Journal of Dermatology Venereology 25: 5-12

Maitra S, Rowland Payne C (2004) The obesity syndrome and *acanthosis nigricans*. Journal of Cosmetic Dermatology 3(4): 202-210

Millington G, Graham-Brown R. Obesity and skin disease (2010) Skin and skin disease throughout life In: Burns A, Breathnach S, Cox N, Griffiths CE, editors Rook's textbook of dermatology 8th ed Oxford: Blackwell publishing

Milman LdM, Grill AB, Müller GP, De Villa D, Souza PRM (2016) Pretibial mucinosis in an euthyroid patient. Anais Brasileiros de Dermatologia 91(1): 100-102

Mir A, Wu T, Orlow SJ (2013) Cutaneous features of Crouzon syndrome with *acanthosis nigricans*. JAMA dermatology 149(6): 737-741

Miraglia E, Visconti B, Bianchini D, Calvieri S, Giustini S (2013) An uncommon association between lipomatous hypertrophy of the interatrial septum (LHIS) and Dercum's disease. European Journal of Dermatology 23(3): 406-407

Misliza A (2009) Sociodemographic and Lifestyle Factors as the risk of diabetic foot ulcer in the University of Malaya Medical Centre. Journal of Health and Translational Medicine 12(1): 15-21

Morales DK, Hogan DA (2010) *Candida albicans* interactions with bacteria in the context of human health and disease. PLoS Pathogens 6(4): e1000886

Mozeika E, Pilmane M, NürNBerG BM, Jemec GB (2013) Tumour necrosis factor-alpha and matrix metalloproteinase-2 are expressed strongly in hidradenitis suppurativa. Acta dermato-venereologica 93(3): 301-304

Nickoloff BJ, Nestle FO (2004) Recent insights into the immunopathogenesis of psoriasis provide new therapeutic opportunities. The Journal of Clinical Investigation 113(12): 1664-1675

Obesity and skin problems (2018). https://www.news-medical.net/health/Obesity-and-skin-problems.aspx; accessed on 22nd December 2018

Ogden CL, Carroll MD, Kit BK, Flegal KM (2014) Prevalence of childhood and adult obesity in the United States, 2011-2012. JAMA 311(8): 806-814

Osman H, Rubeiz N, Tamim H, Nassar AH (2007) Risk factors for the development of *striae gravidarum*. American Journal of Obstetrics and Gynecology 196(1): 62

Otani H (2011) Oxidative stress as pathogenesis of cardiovascular risk associated with metabolic syndrome. Antioxidants & Redox Signaling 15(7): 1911-1926

Parra J, Paye M (2003) EEMCO guidance for the *in vivo* assessment of skin surface pH. Skin Pharmacology and Physiology 16(3): 188-202

Pendrak ML, Yan SS, Roberts DD (2004) Sensing the host environment: recognition of hemoglobin by the pathogenic yeast *Candida albicans*. Archives of biochemistry and biophysics 426(2): 148-156

Potrykus J, Stead D, MacCallum DM, Urgast DS, Raab A, van Rooijen N (2013) Fungal iron availability during deep seated candidiasis is defined by a complex interplay involving systemic and local events. PLoS Pathogens 9(10): e1003676

Prens E, Deckers I (2015) Pathophysiology of hidradenitis suppurativa: an update. Journal of the American Academy of Dermatology 73(5): S8-S11

Rangarajan A, Talora C, Okuyama R, Nicolas M, Mammucari C, Oh H (2001) Notch signaling is a direct determinant of keratinocyte growth arrest and entry into differentiation. The EMBO Journal 20(13): 3427-3436

Rasi A, Soltani-Arabshahi R, Shahbazi N (2007) Skin tag as a cutaneous marker for impaired carbohydrate metabolism: a case-control study. International Journal of Dermatology 46(11): 1155-1159

Ritesh K, Vivian, Argento Y, Amoateng-Adjepong (2013) Obesity-associated abdominal elephantiasis. Case Reports in Medicine (4): 626739

Saad N, Elhadedy K, Ramadan N, Mohmady O, Farid M (2013) The prevalence and risk categorization of diabetic foot complications in cohort group in, Beni Suif, Egypt. Life Science Journal 3: 10

Saffle JR, Morris SE, Edelman L (2008) Fournier's gangrene: management at a regional burn center. Journal of Burn Care & Research 29(1): 196-203

Sari R, Balci MK (2006) Relationship between weight loss and gallbladder motility in obese women. Journal of the National Medical Association 98(10): 1670

Savva A, Kanni T, Damoraki G, Kotsaki A, Giatrakou S, Grech I (2013) Impact of Toll-like receptor-4 and tumour necrosis factor gene polymorphisms in patients with hidradenitis suppurativa. British Journal of Dermatology 168(2): 311-317

Savetsky IL, Torrisi JS, Cuzzone DA, Ghanta S, Albano NJ, Gardenier JC (2014) Obesity increases inflammation and impairs lymphatic function in a mouse model of lymphedema. American Journal of Physiology-Heart and Circulatory Physiology

Sami K, Elshahat A, Moussa M, Abbas A, Mahmoud A (2015) Image analyzer study of the skin in patients with morbid obesity and massive weight loss. Eplasty: 15

Scheinfeld N (2013) Hidradenitis suppurativa: a practical review of possible medical treatments based on over 350 hidradenitis patients. Dermatology Online Journal 19(4)

Schwartz RA (1994) *Acanthosis nigricans*. Journal of the American Academy of Dermatology 31 (1): 1-19

Scholl L, Dörler M, Stücker M (2017) Ulcers in obesity-associated chronic venous insufficiency. Der Hautarzt; Zeitschrift fur Dermatologie, Venerologie, und verwandte Gebiete 68(7): 560-565

Sinha S, Schwartz RA (2007) Juvenile *acanthosis nigricans*. Journal of the American Academy of Dermatology 57(3): 502-508

Smith K, Thiboutot D (2008) Thematic review series: skin lipids. Sebaceous gland lipids: friend or foe? Journal of Lipid Research 49(2): 271-281

Tamega AdA, Aranha AMP, Guiotoku MM, Miot LDB, Miot HA (2010) Association between skin tags and insulin resistance. Anais Brasileiros de Dermatologia 85(1): 25-31

Tung JY, Kiefer AK, Mullins M, Francke U, Eriksson N (2013) Genome-wide association analysis implicates elastic microfibrils in the development of nonsyndromic *striae distensae*. The Journal of Investigative Dermatology 133(11): 2628

Weaver CT, Hatton RD, Mangan PR, Harrington LE (2007) IL-17 family cytokines and the expanding diversity of effector T cell lineages. Annual Review of Immunology 25: 821-852

Weitman ES, Aschen SZ, Farias-Eisner G, Albano N, Cuzzone DA, Ghanta S (2013) Obesity impairs lymphatic fluid transport and dendritic cell migration to lymph nodes. PloS One 8(8): e70703

Wolk K, Wenzel J, Tsaousi A, Witte-Händel E, Babel N, Zelenak C (2017) Lipocalin-2 is expressed by activated granulocytes and keratinocytes in affected skin and reflects disease activity in acne inversa/hidradenitis suppurativa. British Journal of Dermatology 177(5): 1385-1393

Wollina U, Koch A, Heinig B, Kittner T, Nowak A (2013) Acne inversa (Hidradenitis suppurativa): a review with a focus on pathogenesis and treatment. Indian dermatology Online Journal 4(1): 2

Yasuma A, Ochiai T, Azuma M, Nishiyama H, Kikuchi K, Kondo M (2011) Exogenous coproporphyrin III production by *Corynebacterium aurimucosum* and *Microbacterium oxydans* in erythrasma lesions. Journal of medical microbiology 60(7): 1038-1042

Zawieja SD, Wang W, Wu X, Nepiyushchikh Z, Zawieja DC, Muthuchamy M (2011) Impairments in the intrinsic contractility of mesenteric collecting lymphatics in a metabolic syndrome rat model. American Journal of Physiology-Heart and Circulatory Physiology

Zouboulis CC, Desai N, Emtestam L, Hunger R, Ioannides D, Juhász I (2015) European S1 guideline for the treatment of hidradenitis suppurativa/acne inversa. Journal of the European Academy of Dermatology and Venereology 29(4): 619-644

Obesity and Endometriosis

4

Ravindran Jegasothy, Mona M. I. Abdalla, Asita Elengoe, and Methil Kannan Kutty

Abstract

Malaysia has been noted to have the most overweight and obese population among ASEAN countries. Not only in Malaysia, other countries too faced the same conundrum of obesity. Increasing affluence in the population, rapid urbanisation, an increasingly sedentary lifestyle and genetic aspects all influence obesity. Healthcare costs will increase because diseases such as diabetes mellitus (DM), hypertension and heart-related diseases will rise directly in proportion to obesity rates. Obesity is a commonly found phenomenon among women of the reproductive age. Twenty-six percent of women have a body mass index (BMI) of more than 30 kg/m^2 in United Kingdom. There is a preponderance of women by up to three times among admissions to hospitals with a primary diagnosis of obesity. Three times as many female patients seek bariatric surgery as men. The preponderance of the females (up to 70%) will be in the childbearing age. A quarter of women presenting for bariatric surgery suffered from infertility in one study. This book chapter explains whether obesity is linked to endometriosis.

R. Jegasothy
Faculty of Medicine, MAHSA University, Selangor, Malaysia

M. M. I. Abdalla
Human Biology Department, School of Medicine, International Medical University, Kuala Lumpur, Malaysia

A. Elengoe
Biotechnology Department, Faculty of Science, Lincoln University College, Petaling Jaya, Selangor, Malaysia

M. K. Kutty (✉)
Department of Pathology, Faculty of Medicine, Lincoln University College, Petaling Jaya, Selangor, Malaysia

M. K. Kutty, A. Elengoe (eds.), *Obesity and its Impact on Health*,
https://doi.org/10.1007/978-981-33-6408-0_4

Keywords

Obesity · Endometriosis · Women

4.1 Introduction

Worldwide, obesity is a serious public health issue. Obesity means that excess accumulation of lipids in the adipose tissue. A person who has BMI of more than 30 kg/m^2 is considered obese. It affects both genders (male and female). Moreover, it is not only one of the most common health issues in high-income countries such as the United States, United Kingdom but also in poor-income countries like India, Indonesia, etc. The etiologies of obesity are unhealthy food diet, irregular exercises, genetic defects, etc. Fast food (KFC, McDonald, Pizza Hut, Domino, Burger King, etc.) and high calories of fat or sugar food (potato chips, chocolates, ice-cream, soft drinks) cause obesity. Obesity causes chronic diseases such as cardiovascular diseases, hypertension, stroke, kidney-related diseases (e.g. gout), gastrointestinal diseases, respiratory diseases, male and female reproductive system diseases (e.g. polycystic ovarian syndrome (PCOS)) and certain cancers.

This chapter will aim to answer the question of whether obesity is linked to endometriosis. If it is linked it would then be pertinent to ask how this would be so. Next to uterine fibroids, endometriosis is the second most common benign gynaecological condition that is detected in women of the reproductive age group that would be classified as being between 15 and 50 years of age.

Endometriosis is noted when endometrial glands and stroma are found outside the uterine cavity. Endometriosis is usually located within the pelvis but can occur at multiple sites such as the pleural cavity, diaphragm or bowel. It has even been reported in far-flung areas such as the deltoid muscle. While endometriosis is a common and non-malignant process being found in 10–25% of women during diagnostic laparoscopies done for other gynaecological symptoms, ectopic endometrial tissue and the resultant inflammation will cause morbidities such as dyspareunia, dysmenorrhea, chronic pain, and infertility (Mowers et al. 2016). Symptoms can range from being completely asymptomatic, minimal, or severe. The size of the endometrioma does not correlate with the severity of symptoms. Period bleeding from the endometriotic deposit causes a local inflammation and formation of fibrous adhesion. This may lead to the development of an endometrioma or chocolate cyst.

4.2 Role of Adipokines in Pathogenesis and Progression of Endometriosis

A meta-analysis of 11 studies on the relationship between endometriosis and BMI demonstrated that there is an inverse interrelation between BMI and the risk for endometriosis (Liu and Zhang 2017). However, BMI can only be considered as a proxy of adiposity as it reflects the overall body mass rather than adipose tissue mass (Shah and Braverman 2012). Another meta-analysis of 21 publications on the association between endometriosis and the measures of adiposity reported controversial results (Backonja et al. 2016). In two out of five publications that involved bivariate analysis between endometriosis and one or more measures of adiposity, researchers reported a positive association between endometriosis and peripheral adipose tissue below the waist (Backonja et al. 2016). This association could be due to the effect of estrogen in promoting peripheral fat deposition and endometriosis. It is also known as an estrogen-dependent disease (Jasieńska et al. 2004). Endometriosis undergoes resorption in menopause. There are limited research studies on the interrelation between endometriosis and body fat distribution as well as visceral versus subcutaneous deposits of fat which play a significant difference in the outcome and complications associated with obesity (Backonja et al. 2016).

The association between endometriosis and peripheral adiposity can be attributed to the current characterization of the adipose tissue as an endocrine tissue that releases cytokines, chemokines, and adipokines. Adipokines are metabolically-active polypeptides are actively involved in health and diseases (Baranov et al. 2018). In obesity, the hypertrophy of the overloaded adipocytes causes the growth of the adipose tissue that outpaces angiogenesis leading to a hypoxic microenvironment that triggers the release of pro-inflammatory cytokines (Vegiopoulos et al. 2017). In addition to the endocrine function, adipose tissue is also a lymphoid tissue rich in macrophages that tend to be altered to pro-inflammatory cytokines with obesity (Thomas 2017). This inflammatory response in obesity may play a vital role in the pathogenesis and progression of endometriosis which is a chronic inflammatory disease (Baranov et al. 2018). Expansion of adipose tissue is also related to alterations in the adipokines secreted including leptin, adiponectin, resistin, visfatin, omentin and vaspin (Guerrero-García et al. 2016). The alterations in those adipokines associated with obesity may also play a significant role in the pathogenesis and progression of endometriosis.

Leptin is an adipokine that is produced predominantly by the adipose tissue and secreted into the circulation. It plays a crucial role in energy homeostasis by promoting satiety and energy expenditure (Friedman 2016). Leptin level is positively correlated with adipose tissue accumulation, especially with subcutaneous adiposity (van Harmelen et al. 1998). Leptin is an important immune response modulator and a high level of it modulates the immune response to a pro-inflammatory profile (Scotece et al. 2017).

The available data on the interrelationship between leptin and endometriosis enforces its role in the pathogenesis of endometriosis. Studies have shown an increase of leptin level in the peritoneal fluid (PF) of patients with endometriosis (Pandey et al. 2010; Malhotra et al. 2011; Liu and Zhang 2017) and an increased expression rate of leptin and leptin receptors (ObR) in endometrial tissue samples from ovarian endometriomas (Choi et al. 2013) as well as deep infiltrating endometriosis (Gonçalves et al. 2015) as compared with normal endometrium.

Leptin was found to promote the growth, migration, and invasion of the endometrium (Oh et al. 2013; Ahn et al. 2015) which are important features for endometriosis. Based on Malhotra et al. (2011) study, it has also been demonstrated that the increase of leptin levels in the PF of patients with endometriosis was associated with an increase of IL-8, an inflammatory cytokine that plays a significant role in the pathogenesis of endometriosis. Leptin can be used as a marker for the severity, stages, and progression of endometriosis as evidenced by the significant increase of leptin level during the early stages of endometriosis; stages 1 and 2 as compared with the advanced stages; stages 3 and 4 (Malhotra et al. 2011). Research studies have shown higher levels of leptin in infertile women with endometriosis as compared with infertile women without endometriosis. This reflects on the implication of using leptin as a marker for infertility; a common complication of endometriosis.

Adiponectin is the most abundant adipokine in circulation and it is exclusively produced by the adipose tissue. Adiponectin actions are mediated by AdipoR 1 and AdipoR 2 transmembrane adiponectin receptors. Cho et al. (2017) study has been shown that adiponectin level is inversely associated with obesity and insulin resistance. The loss of the protective effects of adiponectin in obese individuals was found to be correlated with obesity-related complications such as diabetes mellitus, metabolic syndrome (Cho et al. 2017) and obesity-related malignancies such as ovarian cancer and colorectal cancers (Tworoger and Huang 2016; Li et al. 2017; Otani et al. 2017).

Adiponectin inhibits the synthesis of IL-8 induced by TNF-alpha (Kobashi et al. 2005). Consequently, the decreased level of adiponectin causes an increase of IL-8 that is known to play a vital role in the pathogenesis of endometriosis (Malhotra et al. 2011). The expression of the two receptors of adiponectin was shown to be reduced in women with implantation failure as compared with fertile women (Dos Santos et al. 2012). This implies the importance of adiponectin in uterine receptivity and as a regulator for the reproductive system (Dobrzyn et al. 2018).

The low adiponectin concentration in the PF in women with endometriosis and its further drop with the advanced stages of the endometriosis confirms the importance of adiponectin as an anti-endometriotic factor (Takemura et al. 2005, 2006). Additionally to the anti-inflammatory and anti-angiogenic effects of adiponectin, research studies have shown that adiponectin inhibits endometriosis by decreasing the viability and proliferation of human endometrial stromal cells and increasing the expression of Adio R1 and Adipo R 2 expression in the endometrial stromal cells (Bohlouli et al. 2013). These data reflect the possible use of adiponectin as a marker for the severity of endometriosis and also as a treatment option.

Resistin is another adipokine that is synthesized and released by adipose tissue and other cells such as skeletal muscle, adrenal gland, intestinal epithelium and pituitary gland (Guerrero-García et al. 2016). Resistin is a known mediator for the development of insulin resistance, diabetes mellitus type 2 and heart-related diseases (Cymbaluk-Płoska et al. 2016; Cabrera de León et al. 2014; Haluzik et al. 2006). Research studies have shown higher levels of resistin in females as compared with males (Cabrera de León et al. 2014). Higher levels of resistin were also reported among obese subjects as compared with lean subjects (Nieva-Vazquez et al. 2014). The higher levels of resistin were shown to be positively correlated with some inflammatory mediators, such as TNF-alpha, C-reactive protein (CRP) and IL- 1β (Nieva-Vazquez et al. 2014).

Research studies have been demonstrated that the serum and PF levels of resistin were higher in patients with endometriosis as compared with women without endometriosis (Nieva-Vazquez et al. 2014; Yi et al. 2010). The increase in serum and PF of resistin levels related to endometriosis may be due to the enhanced expression rates of resistin mRNA in the ectopic endometrial tissues as compared with normal endometrium (Oh et al. 2017).

Recently, Zhang et al. (2017) investigated the outcome of *in vitro* fertilization/embryo transfer (IVF-ET) and its relationship with serum and follicular levels of resistin and interleukin-23 (IL-23) in two groups of infertile women with endometriosis; one group in stages I and II of endometriosis versus another group in stages III and IV of the disease. They have found a significantly higher level of resistin and (IL-23) in both serum and follicular fluid of the group in the advanced stages of endometriosis as compared with the early stages group. That was accompanied by the failure of IVF-ET due to associated inflammation and immune rejection caused by the increase of IL-23, an inflammatory cytokine that promotes inflammation and participates in autoimmune diseases development (Zhang et al. 2017).

Omentin, visfatin and vaspin are relatively recently discovered adipokines. Omentin and vaspin are released mainly from visceral fat (Guerrero-García et al. 2016). Omentin has many protective effects such as anti-diabetic, anti-atherogenic and anti-inflammatory effects (Yang et al. 2006). Omentin level decreases with obesity and this may play a vital role in the development of the inflammatory condition and its related complications (de Souza Batista et al. 2007). The decreased level of omentin in obese women may play a role in the pathogenesis of endometriosis. Omentin may be of a potential therapeutic effect for inflammatory diseases such as endometriosis. Contrary to omentin, vaspin level increases with obesity and obesity-associated insulin resistance. Visfatin adipokine is known to promote the production of inflammatory mediators and a current study reported a significant positive interrelation between the serum level of visfatin and the poor prognosis of endometrial cancer (Moschen et al. 2007). Increased levels of vaspin and visfatin may play a vital role in the pathogenesis of endometriosis.

4.3 Medical Treatment Options and Considerations

Medical procedures do not improve fertility, reduce endometriomas, or treat complications of deep endometriosis such as ureteral obstruction. Women with these problems should proceed with therapy targeted at a specific problem (Table 4.1) (Dunselman et al. 2014; Shaw et al. 2005).

Obesity affects the pharmacokinetics and pharmacodynamics of many drugs. It must be taken into consideration when considering drug therapy for endometriosis. Variations in drug absorption and storage may also contribute to the side effects when on these drugs which often have to be given for the duration of 6–9 months or even more on some occasions.

Table 4.1 The effects of hormones on endometrium

Hormones/drugs	Endometrial change	Effect on ectopic endometrium
Oestrogens	Proliferation/hyperplasia	Proliferation/hyperplasia
Drugs causing pseudomenopause GnRH analogues Oophorectomy	Atrophy	Atrophy, resorption
Progestogens Pseudopregnancy	Secretory change Decidualisation	Secretory change, resorption
Androgens Danazol Gestrinone	Atrophy	Atrophy, regression

Combined oral contraceptives for the management of endometriosis would not be advisable in obese patients due to a high risk of thromboembolism.

Oral progestogens exert an anti-oestrogenic effect. The treatment over a period of 6–9 months produces a state of pseudopregnancy which ultimately leads to regressions and resorption of the disease. Weight gain and irregular bleeding are side effects of progestogens. These side effects would also be of concern in obese patients as a higher dosage would be required because of the boy weight. Fertility can be impaired for up to 2 years after prolonged hormonal therapy. Mirena IUCD which releases a locally acting progestogen has minimal side effects.

Danazol, a synthetic derivative of ethinyl testosterone, inhibits pituitary gonadotrophins. It is mildly anabolic, antioestrogenic and antiprogestational. It induces a pseudomenopausal state. The lesions will regress and there are more side effects with this medication. Of particular concern to obese patients is weight gain, excessive sweating and muscle cramps.

Gestrinone is a 19-nortestosterone derivative similar in action to danazol but has fewer side effects and is long-acting.

Gonadotrophin releasing hormone (GnRH) is administered continuously to down-regulate and suppress pituitary gonadotrophins and cause atrophy of endometriotic tissues in 90% of cases. It does not cause weight gain issues.

Aromatase inhibitors like letrozole and anastrozole are also used. They are anti-oestrogen and should be given with Vitamin D and calcium to prevent osteoporosis. There are no weight-related issues.

4.3.1 Surgical Options

Laparoscopy has played an increasing role in the management of endometriosis. This offers the advantage of conserving the ovaries, fallopian tubes and improving fertility. Aspiration of peritonel fluid can be performed to reduce dysmenorrhoea and pelvic pain as this procedure removes PGE2 with the fluid.

Small lesions of less than 3 cm can be diathermised. Larger lesions can be excised. Cystectomy can be done for pain relief. This is done usually with adhesiolysis. Laser uterosacral nerve ablation can be done in some cases.

Recurrence following conservative surgery is 10% at the end of 1 year and 25% at the end of 3 years. Adhesions form in 10%.

Laparotomy is required in advanced stages and larger lesions. Dissection and excision of chocolate cysts, salpingo-oophorectomy and abdominal hysterectomy with bilateral salpingo-oohorectomy is usually performed through a laparotomy. Surgery can be difficult due to adhesions.

Surgery in obese patients carries its challenges such as inadequate access and visualisation of diseased tissues due to a deep pelvis and thick abdominal wall. There will also be an anaesthetic difficulty as well as post-operative complications due to respiratory complications arising from a heavy chest wall as well as delays in mobilising the patient.

4.4 Clinical Trials for Obese Patients with Endometriosis Cancer (EC)

Based on Schuler et al. (2015) study, it has been demonstrated that metformin decreased the anti-tumour activity in obese patients with endometrial cancer (EC). It reduced anti-tumour activity via inhibition of the mTOR signalling pathway. The target proteins involved in the mTOR signalling pathway were determined using the immunohistochemistry method. It interferes with the metabolism of glucose and fat.

Bijen et al. (2011) study, revealed that total laparoscopic hysterectomy (TLH) was a better procedure compared to total abdominal hysterectomy (TAH) for obese patients with EC. It was safe and low cost.

Based on von Gruenigen et al. (2006) study, it has been reported that BMI is closely associated with obese patients with EC. The results show that a decreased in the survival of patients who have BMI (more than 40 kg/m^2) when compared to patients (less than 30 kg/m^2).

Foley and Lee (1990) carried out an analysis of acute surgical complications in obese patients with EC and normal patients. They found that obese people had higher chances of surgical complications such as wound infection and wound dehiscence. Obese patients pass through a longer operation period. Moreover, they have to stay at the hospital for a longer time. They also suffered from a high loss of blood.

4.5 Conclusion

Endometriosis is a chronic disease that causes morbidity by manifesting as chronic pelvic pain, dyspareunia, infertility and quality of life. Obesity plays a vital role in the causation of endometriosis due to the role of adipokines in the adipose tissue as eluded to above. Obesity also has an indirect role in morbidity arising as a result of errors from diagnostic difficulties in imaging due to a thick abdominal wall, side effects due to higher doses of medication as well as even more weight gain and surgical difficulties compounded by anaesthetic complications that manifest in the obese individual as well as access difficulties during the procedure itself. Malignancy is rare (1:150) and mostly manifests as endometroid cancer, a form of epithelial ovarian cancer. A high index of suspicion must be maintained in obese individuals for malignant change due to adipose tissue contributing to higher endogenous oestrogen levels. People should aware the seriousness of obesity. They should take healthy food (e.g. vegetables, fruits and whole grains); drink a lot of water; avoid junk food; regular exercise or physical activity (e.g. walking, jogging, swimming, cycling, etc.) to prevent the obesity problem.

References

Ahn JH, Choi YS, Choi JH (2015) Leptin promotes human endometriotic cell migration and invasion by up-regulating MMP-2 through the JAK2/STAT3 signaling pathway. Mol Hum Reprod 21(10): 792-802

Backonja U, Buck Louis GM, Lauver DR (2016) Overall adiposity, adipose tissue distribution, and endometriosis: A systematic review. Nurs Res 65(2): 151-166

Baranov V, Malysheva O, Yarmolinskaya M (2018) Pathogenomics of endometriosis development. Int J Mol Sci 19(7): E1852

Bijen CB, de Bock GH, Vermeulen KM, Arts HJ, ter Brugge HG, van der Sijde R, Kraayenbrink AA, Bongers MY, van der Zee AG, Mourits MJ (2011) Laparoscopic hysterectomy is preferred over laparotomy in early endometrial cancer patients, however not cost effective in the very obese. Eur J Cancer 47(14): 2158-2165

Bohlouli S, Khazaei M, Teshfam M, Hassanpour H (2013) Adiponectin effect on the viability of human endometrial stromal cells and mRNA expression of adiponectin receptors. Int J Fertil Steril 7(1): 43-48

Cabrera de León A, Almeida González D, González Hernández A, Juan Alemán Sánchez J, Brito Díaz B, Domínguez Coello S, Marcelino Rodríguez I, Gregorio Oliva García J, Aguirre Jaime A, Rodríguez Pérez Mdel C (2014) The association of resistin with coronary disease in the general population J Atheroscler Thromb 21(3): 273-281

Cho SA, Joo HJ, Cho JY, Lee SH, Park JH, Hong SJ, Yu CW, Lim DS (2017) Visceral fat area and serum adiponectin level predict the development of metabolic syndrome in a community-based asymptomatic population. PLoS One 12(1): e0169289

Choi YS, Oh HK, Choi JH (2013) Expression of adiponectin, leptin, and their receptors in ovarian endometrioma. Fertil Steril 100(1): 135-141

Cymbaluk-Płoska A, Chudecka-Głaz A, Pius-Sadowska E, Sompolska-Rzechuła A, Machaliński B, Menkiszak J (2016) Association between resistin levels and cardiovascular disease events in older adults: the health, aging and body composition study. Atherosclerosis 245: 181-186

de Souza Batista CM, Yang RZ, Lee MJ, Glynn NM, Yu DZ, Pray J, Ndubuizu K, Patil S, Schwartz A, Kligman M, Fried SK, Gong DW, Shuldiner AR, Pollin TI, McLenithan JC (2007) Omentin plasma levels and gene expression are decreased in obesity. Diabetes 56(6): 1655-1661

Dobrzyn K, Smolinska N, Kiezun M, Szeszko K, Rytelewska E, Kisielewska K, Gudelska M, Kaminski T (2018) Adiponectin: A new regulator of female reproductive system. Int J Endocrinol 2018: 7965071

Dos Santos E, Serazin V, Morvan C, Torre A, Wainer R, de Mazancourt P, Dieudonné MN (2012) Adiponectin and leptin systems in human endometrium during window of implantation. Fertil Steril 97(3): 771-778

Dunselman GA, Vermeulen N, Becker C, Calhaz-Jorge C, D'Hhooghe T, De Bie B, Heikinheimo O, Horne AW, Kiesel L, Nap A, Prentice A, Saridogan E, Soriano D, Nelen W

(2014) ESHRE guideline: management of women with endometriosis. Hum Reprod 29(3): 400-412

Guerrero-García JJ, Carrera-Quintanar L, López-Roa RI, Márquez-Aguirre AL, Rojas-Mayorquín AE, Ortuño-Sahagún D (2016) Multiple sclerosis and obesity: Possible roles of adipokines. Mediators Inflamm 2016: 4036232

Friedman J (2016) The long road to leptin. J Clin Invest 126: 4727-4734

Foley K, Lee RB (1990) Surgical complications of obese patients with endometrial carcinoma. Gynecol Oncol 39(2): 171-174

Gonçalves HF, Zendron C, Cavalcante FS, Aiceles V, Oliveira MA, Manaia JH, Babinski MA, Ramos CF (2015) Leptin, its receptor and aromatase expression in deep infiltrating endometriosis. J Ovarian Res 8: 53

Haluzik and D. Haluzikova (2006) The role of resistin in obesity-induced insulin resistance. Current Opinion in Investigational Drugs 7(4): 306-311

Jasieńska G, Ziomkiewicz A, Ellison PT, Lipson SF, Thune I (2004) Large breasts and narrow waists indicate high reproductive potential in women. Proc Biol Sci 271(1545): 1213-1217

Kobashi C, Urakaze M, Kishida M, Kibayashi E, Kobayashi H, Kihara S, Funahashi T, Takata M, Temaru R, Sato A, Yamazaki K, Nakamura N, Kobayashi M (2005) Adiponectin inhibits endothelial synthesis of interleukin-8. Circ Res 97(12): 1245-1252

Li X, Yu Z, Fang L, Liu F, Jiang K (2017) Expression of adiponectin receptor-1 and prognosis of epithelial ovarian cancer patients. Med Sci Monit 23: 1514-1521

Liu Y, Zhang W (2017) Association between body mass index and endometriosis risk: a meta-analysis. Oncotarget 8(29): 46928-46936

Malhotra N, Karmakar D, Tripathi V, Luthra K, Kumar S (2011) Correlation of angiogenic cytokines-leptin and IL-8 in stage, type and presentation of endometriosis. Gynecol Endocrinol 28(3): 224-247

Moschen AR, Kaser A, Enrich B, Mosheimer B, Theurl M, Niederegger H, Tilg H. Visfatin (2007) An adipocytokine with proinflammatory and immunomodulating properties. J Immunol 178(3): 1748-1758

Mowers EL, Lim CS, Skinner B, Mahnert N, Kamdar N, Morgan DM, As-Sanie S (2016) Prevalence of endometriosis during abdominal or laparoscopic hysterectomy for chronic pelvic pain. Obstet Gynecol 127(6): 1045

Nieva-Vazquez A, Pérez-Fuentes R, Torres-Rasgado E, López-López JG, Romero JR (2014) Serum resistin levels are associated with adiposity and insulin sensitivity in obese Hispanic subjects. Metab Syndr Relat Disord 12(2): 143-148

Oh HK, Choi YS, Yang YI, Kim JH, Leung PC, Choi JH (2013) Leptin receptor is induced in endometriosis and leptin stimulates the growth of endometriotic epithelial cells through the JAK2/STAT3 and ERK pathways. Mol Hum Reprod 19(3): 160-168

Oh YK, Ha YR, Yi KW, Park HT, Shin JH, Kim T, Hur JY (2017) Increased expression of resistin in ectopic endometrial tissue of women with endometriosis. Am J Reprod Immunol 78(5): 12726

Otani K, Ishihara S, Yamaguchi H, Murono K, Yasuda K, Nishikawa T, Tanaka T, Kiyomatsu T, Hata K, Kawai K, Nozawa H, Watanabe T (2017) Adiponectin and colorectal cancer. Surg Today 47(2): 151-158

Pandey N, Kriplani A, Yadav RK, Lyngdoh BT, Mahapatra SC (2010) Peritoneal fluid leptin levels are increased but adiponectin levels are not changed in infertile patients with pelvic endometriosis. Gynecol Endocrinol 26(11): 843-849

Scotece M, Pérez T, Conde J, Abella V, López V, Pino J, Gonzalez-Gay MA, Gomez-Reino JJ, Mera A, Gomez R, Gualillo O (2017) Adipokines induce pro-inflammatory factors in activated Cd^{4+} T cells from osteoarthritis patient. J Orthop Res 35(6): 1299-1303

Schuler KM, Rambally BS, DiFurio MJ, Sampey BP, Gehrig PA, Makowski L, Bae-Jump VL (2015) Antiproliferative and metabolic effects of metformin in a preoperative window clinical trial for endometrial cancer. Cancer Med 4(2): 161-173

Shah NR, Braverman ER (2012) Measuring adiposity in patients: the utility of body mass index (BMI), percent body fat, and leptin. PLoS One 7(4): e33308

Shaw KA, O'Rourke P, Del Mar C, Kenardy J (2005) Psychological interventions for overweight or obesity. Cochrane Database Syst Rev (2): CD003818

Takemura Y, Osuga Y, Harada M, Hirata T, Koga K, Yoshino O, Hirota Y, Morimoto C, Yano T, Taketani Y (2005) Concentration of adiponectin in peritoneal fluid is decreased in women with endometriosis. Am J Reprod Immunol 54(4): 217-221

Takemura Y, Osuga Y, Yamauchi T, Kobayashi M, Harada M, Hirata T, Morimoto C, Hirota Y, Yoshino O, Koga K, Yano T, Kadowaki T, Taketani Y (2006) Expression of adiponectin receptors and its possible implication in the human endometrium. Endocrinol 147(7): 3203-3210

Thomas D, Apovian C (2017) Macrophage functions in lean and obese adipose tissue. Metabolism 72: 120-143

Tworoger SS, Huang T (2016) Obesity and ovarian cancer. Recent Results Cancer Res 208: 155-176

van Harmelen V, Reynisdottir S, Eriksson P, Thörne A, Hoffstedt J, Lönnqvist F, Arner P (1998) Leptin secretion from subcutaneous and visceral adipose tissue in women. Diabetes 47(6): 913-917

Vegiopoulos A, Rohm M, Herzig S (2017) Adipose tissue: between the extremes. EMBO J 36(14): 1999-2017

von Gruenigen VE, Tian C, Frasure H, Waggoner S, Keys H, Barakat RR (2006) Treatment effects, disease recurrence, and survival in obese women with early endometrial carcinoma: a Gynecologic Oncology Group study. Cancer 107(12): 2786-2791

Yang RZ, Lee MJ, Hu H, Pray J, Wu HB, Hansen BC, Shuldiner AR, Fried SK, McLenithan JC, Gong DW (2006) Identification of omentin as a novel depot-specific adipokine in human adipose tissue: possible role in modulating insulin action Am J Physiol Endocrinol Metab 290 (6): E1253-1261

Yi KW, Shin JH, Park HT, Kim T, Kim SH, Hur JY (2010) Resistin concentration is increased in the peritoneal fluid of women with endometriosis. Am J Reprod Immunol 64(5): 318-323

Zhang QF, Chen GY, Liu Y, Huang HJ, Song YF (2017) Relationship between resistin and IL-23 levels in follicular fluid in infertile patients with endometriosis undergoing IVF-ET. Adv Clin Exp Med 26(9): 1431-1435

Hypogonadism

5

Prameela Kannan Kutty and Methil Kannan Kutty

Abstract

Obesity per se is associated with various hormonal derangements and a pro-inflammatory state. Obese patients may also present with hypogonadism either as part of the obese state itself or as an additional or distinguishable clinical feature. As such, when dealing with these two issues, there is debate as to their interlinks. This chapter gives the reader an outline of the categories of hypogonadism and, in some instances, their related links.

Keywords

Obesity · Pro-inflammatory · Hypogonadism

5.1 Introduction

The term hypogonadism typically denotes permanent instead of transitory or revocable faults and generally proposes deficits in the levels of reproductive hormones and at times could also be associated with problems involving fertility. Obesity impacts all organ systems including the reproductive system. Hence an understanding of the subject of hypogonadism is pertinent to this topic. In boys, hypogonadism may cause an absence of secondary sexual development or failure of normal progression to puberty, whereas in girls, hypogonadism may be linked to absence

P. K. Kutty
Department of Paediatrics, National Defence University of Malaysia, Kuala Lumpur, Malaysia

M. K. Kutty (✉)
Department of Pathology, Faculty of Medicine, Lincoln University College, Petaling Jaya, Selangor, Malaysia

M. K. Kutty, A. Elengoe (eds.), *Obesity and its Impact on Health*,
https://doi.org/10.1007/978-981-33-6408-0_5

of pubertal development and progression or with menstrual irregularities (Viswanathan and Eugster 2011).

Hypogonadotropic hypogonadism is due to anomalies that occur within the hypothalamus or pituitary, whereas primary gonadal failure is characterized as hypergonadotropic hypogonadism (Viswanathan and Eugster et al. 2011).

There is a clear association between the obese state and hypogonadism (Rao et al. 2013). Studies in populations indicate that obesity a very important factor resulting in testosterone deficiency (Tajar et al. 2010).

Similarly, testosterone deficiency can cause increased adipogenesis and visceral obesity as observed in some conditions (Tsai et al. 2000).

5.2 Classification of Hypogonadism (Kumar et al. 2010)

Primary hypogonadism implies that the problem is in the testis or ovaries. The low levels of testosterone estrogen stimulate feedback control in the hypothalamus and pituitary gland resulting in elevated LH and FSH levels. Hence this is a "peripheral" end-organ abnormality of hormone production.

In secondary hypogonadism, the testes are normal but do not function normally. Both LH and FSH are normal or low, so here there is central hypothalamic or pituitary involvement. The testis and ovaries are normal.

At other times, end organs like the testes may not respond to the hormones produced in which case the problem is one of end-organ receptor insensitivity.

5.3 Primary Hypogonadism

Primary hypogonadism where LH and FSH are typically high. Examples of genetic conditions include Turner's Syndrome and Klinefelter's Syndrome, Noonan's Syndrome, XY with XRY gene immunity. It may be acquired as part of mumps orchitis or a varicocele of the testis could reduce hormone production. A varicocele can reduce the production of testosterone (Çayan et al. 2020).

5.4 Secondary Hypogonadism

The hypothalamus produces gonadotrophin-releasing hormone (GRH), this stimulates the pituitary to produce FSH and LH and the testis then produces testosterone. When both LH and FSH are low, the problem is central in the hypothalamus or pituitary gland. Examples of hypothalamic defects causing hypogonadism occur in Kallmann syndrome (Zhang et al. 2020), pituitary defects can occur in hypopituitarism and pituitary hypoplasia.

5.5 End Organ Insensitivity

Androgens produced by the testes exert their actions through receptors that bind to testosterone. Androgen insensitivity (Lanciotti et al. 2019) due to lack of response to the hormone as a result of receptor deficiency cause then causes different sexual characteristics. These phenotypes differ from the chromosomal sex of the individual.

5.6 Causes of the Different Types of Hypogonadism

Primary causes, secondary causes, or problems associated with end-organ insensitivity can be caused by a spectrum of inherited or acquired causes.

Klinefelter's syndrome (Klinefelter et al. 1942) is a chromosomal disorder affecting males with a spectrum of differences in clinical severity. It affects both physical and intellectual function. This syndrome is associated with taller than average somatic growth and many individuals are infertile. This condition is a result of a series of anomalies of the sex chromosomes. In Klinefelter's Syndrome, there are extra X chromosomes in addition to the Y chromosome, whereas in a normal male there is one X and one Y chromosome. Mosaic Klinefelter syndrome (46, XY/47, XXY) which is not inherited. It happens as a random error during cell division early in fetal development. Consequently, some of the body's cells have the usual one X chromosome and one Y chromosome (46, XY), and other cells have an extra copy of the X chromosome (47, XXY) (Visootsak and Graham 2006).

In Noonan syndrome, there are clinical similarities to Turner's syndrome. The male has short stature with characteristic facial features. Heart defects, bleeding problems, developmental delays and skeletal malformations occur. Cryptorchidism is present in about 1 in 2000 people with the condition (Balasubramanian and Crowley 2017; Vikas and Maximilian 2014).

Recurrence in families suggests an autosomal dominant mode of inheritance but most cases occur as a result of new mutations. Genetic alterations of PTPN11, KRAS, SOS1, RAF 1, have been described. The diagnosis of Noonan's Syndrome is both by clinical appearance and molecular genetic testing (Liao and Mehta 2019).

5.7 Acquired Causes of Hypogonadism

Drug-induced

1. Androgen deficiency associated with opioid usage
   - Codeine, morphine, fentanyl, methadone and others (Rubinstein and Carpenter 2017)
   - Gonadal hormones are crucial factors that can modulate the pain experience; and with the usage of these drugs, concentrations of testosterone fall within a few hours of opioid use (Rubinstein et al. 2013).
2. Hypogonadism associated with anabolic steroids or anabolic steroid-induced

hypogonadism

- Symptomatic hypogonadism can be caused by anabolic-androgen steroids and contingent on the type of anabolic steroid used the dose and the duration. This condition is more frequently seen in young men (Coward et al. 2013).

5.8 Anabolic Steroid-Induced Hypogonadism (ASIH)

ASIH subdues the hypothalamic-pituitary axis. There is a reduced release of luteinising hormone and follicle-stimulating hormone. Clinical symptoms include testicular atrophy, gynecomastia, sexual dysfunction and others (Coward et al. 2013; Rahnema et al. 2014; Hyun 2018).

5.9 Infections Causing Hypogonadism

Mumps parotitis can cause and if so, in adolescents and adults will result could in testicular damage with the interference of the production of testosterone. An increase in mumps orchitis in pubertal males has been linked to a reduced uptake of the measles-mumps-rubella (MMR) vaccine in1990s in children who are now teenagers. The extra-salivary complications of mumps orchitis, as a complication of the mumps virus, can result in testicular atrophy, infertility or reduced fertility (Davis et al. 2010).

- Prenatal causes of hypogonadism
  Prenatal exposures such as to the endocrine disrupter diethylstibesterol for miscarriage (Reed and Fenton 2013).
- Undescended testis
  In utero the testes develop within the abdomen and later on in gestation, they descend to the perineum. If this does not happen, it usually corrects itself but if left uncorrected it may cause testicular malfunction. This results in reduced production of testosterone and hypogonadism (Rodprasert et al. 2019).
- Trauma
  The testes are unprotected by the abdomen so it is prone to trauma. If both testes are involved, it could affect testosterone production.
- Traumatic brain injury
  Injury to the pituitary gland or the hypothalamus is a recognised caused of hypogonadotropic hypogonadism (Dandona and Dhinsa 2011). This may be due to surgery, injury, infection radiation and so on. Persistent Hypogonadotropic Hypogonadism status in men predicts poor outcome after severe traumatic brain injury (Barton et al. 2016)
- Hemochromatosis
  Excess blood iron can be deposited in the pituitary gland. Excesses can occur as a result of an inherited anomaly, or as a result of treatment in specific conditions

such as thallassaemia without timely iron chelation therapy. This causes hypogonadism when the iron is deposited in the pituitary gland or the testes. Hypogonadotropic hypogonadism (HH) is a complications of hemochromatosis (Osta et al. 2017)

- Cancer treatment (Garcia et al. 2006; Burney and Garcia 2012)
  Although a reversible cause of hypogonadism, about 30% of men on cancer chemotherapy can have a testicular failure. Many have a germinal epithelial failure. Sperm preservation before initiating anti-cancer therapy is an option in these circumstances.
- Normal aging process
  Up to 30% of older men have a testosterone level that is below normal and the prerequisite for treatment in these circumstances is debatable (Boehm et al. 2015).

5.10 Syndromes Associated with Secondary Hypogonadism

The commonest cause of male hypogonadism in adults is obesity (Niederberger 2014). However, male obesity-associated secondary hypogonadism (MOSH) is a diagnosis of exclusion, which is made after the primary causes of hypogonadism have been ruled out (Mushannen et al. 2019).

Kallman Syndrome is due to the partial or complete deficiency in the olfactory bulb in the brain.

Kallman syndrome is a condition that interferes with the attainment of puberty. It is typically associated with anosmia or hyposmia. If untreated, individuals have poorly developed secondary sexual characteristics and are often infertile. They are at higher risk of developing osteoporosis with facial and skeletal derangements (Balasubramanian and Crowley 2017).

A decreased level of sex steroids in males and females is due to the failure of the central production of the gonadotrophin-releasing hormone. Kisspeptin is a hypothalamic hormone that modifies GnRH secretion and is implicated in puberty regulation. Kisspetin interacts with neuropepetides to regulate GnRH generation. Loss of kisspeptin signaling causes hypogonadotrophic hypogonadism (Clarke et al. 2015).

Pituitary gland disorders affect hormone release to the testicles and impact the normal production of testosterone. Central tumours such as pituitary gland tumours or surgical or radiotherapy for brain tumors may cause hypogonadism by impairing pituitary function. Inflammatory disorders such as tuberculosis and histiocytosis involve the hypothalamus and pituitary gland and affect testosterone production, causing hypogonadism.

The Human immunodeficiency virus (HIV) and Acquired Immune Deficiency Syndrome (AIDS) can impact the brain or the testes (Wong et al. 2017) and affect related hormones and releasing factors. Drug effects are considered in the differential diagnosis of many clinical conditions particularly when the clinical history is suggestive or due to an index of suspicion in the particular case. The opioid drug class can cause a deficiency of gonadotrophins with reduced sex steroids. As the dismal aftermath of the increased use of longstanding opioid therapy, there has been an escalated incidence of opioid linked endocrinopathy or opioid associated androgen deficiency (OPIAD)(Gudin et al. 2015; Smith and Elliott 2012).

Patients may experience various sexual problems, tiredness, hot flushes with suggestive clues in the physical examination such as anemia, osteopenia and so on. Medications that potentially interfere with endocrine function can cause hypogonadism in the fetus when a pregnant mother consumes such medication.

Traumatic brain injury (TBI) has been known to cause hypogonadism in a number of survivors. This can result in both acutely, as well as a long term problem after brain injury. Obesity is a cause of hypogonadism. This may be a result of the interplay of hormones, adipokines, metabolic disturbances and other derangements in obesity. Syndromic associations include Bardet-Biedl syndrome, Prader-Willi Syndrome and various other syndromes (Bardet Biedl Syndrome Foundation 2019; Prader-Willi Syndrome 2020).

5.11 Conclusion

It is clear that hypogonadism is a problem that must be identified and addressed in a holistic manner because of its various causes and pathologies. The different etiopathogenesis of hypogonadism is such that when dealing with obesity, the history pertaining to sexual maturity and the examination of the genitalia is paramount. It is of note that male hypogonadism is commonly recognised in patients with obesity. Different pathophysiological mechanisms may include increased inflammation due to increased fat, enhanced conversion of testosterone to estrone which impacts the hypothalamic-pituitary gonadal axis and delayed development of secondary sexual characteristics in adolescent males. Obesity and hypogonadism is a narrative review highlighting the need for high-quality data in adolescents (Mushannen et al. 2019).

Reflecting on the other causes of hypogonadism including syndromic associations, and various congenital as well as acquired causes, an index of suspicion coupled with a thorough history, physical examination and investigative profile is required to clinch the diagnoses.

It is crucial to also note that low testosterone levels are not only linked to sexual health but also cardiovascular diseases, metabolic diseases, insulin sensitivity and decreased bone density.

References

Balasubramanian R, Crowley Jr WF (2017) Isolated gonadotropin-releasing hormone (GnRH) deficiency. InGeneReviews®. University of Washington, Seattle

Bardet Biedl Syndrome Foundation. [(accessed on 12 December 2020)];2019 Available online: https://www.bardetbiedl.org/what-is-bbs/

Barton DJ, Kumar RG, McCullough EH , Galang G, Arenth PM, Berga SL, Wagner AK (2016) Persistent Hypogonadotropic Hypogonadism in Men After Severe Traumatic Brain Injury. Journal of Head Trauma Rehabilitation 31 (4):277-287

Burney BO, Garcia JM (2012) Hypogonadism in male cancer patients. Journal of Cachexia, Sarcopenia and Muscle 3 (3):149-155

Boehm U, Bouloux PM, Dattani MT, de Roux N, Dodé C, Dunkel L, Dwyer AA (2015) Expert consensus document: European consensus statement on congenital hypogonadotropic hypogonadism-pathogenesis, diagnosis and treatment. Nature Reviews Endocrinology 11 (9): 54764

Çayan S, Akbay E, Saylam B, Kadıoğlu A (2020) Effect of Varicocele and Its Treatment on Testosterone in Hypogonadal Men with Varicocele: Review of the Literature. Balkan Medical Journal 37 (3):121-124

Clarke H, Dhillo WS, Jayasena CN (2015) Comprehensive Review on Kisspeptin and Its Role in Reproductive Disorders. Endocrinology and Metabolism 30 (2):124

Coward RM, Rajanahally S, Kovac JR, Smith RP, Pastuszak AW, Lipshultz LI (2013) Anabolic steroid induced hypogonadism in young men. Journal of Urology 190: 2200-2205

Dandona P, Dhindsa S (2011) Update: Hypogonadotropic Hypogonadism in Type 2 Diabetes and Obesity. The Journal of Clinical Endocrinology & Metabolism 96 (9):2643-2651

Davis NF, McGuire BB, Mahon JA, Smyth AE, O'Malley KJ, Fitzpatrick JM, (2010) The increasing incidence of mumps orchitis: a comprehensive review. BJU International 105 (8):1060-1065

Garcia JM, Li H, Mann D, Epner D, Hayes TG, Marcelli M, Cunningham GR (2006) Hypogonadism in male patients with cancer. Cancer 106 (12):2583-2591

Gudin JA, Laitman A, Nalamachu S(2015) Opioid Related Endocrinopathy: Table 1. Pain Medicine 16 (suppl 1):S9-S15

Hyun JP (2018) Anabolic steroid-induced hypogonadism: a challenge for clinicians. Journal of Exercise Rehabilitation 14(1): 2-3

Klinefelter HF, Reifenstein EC, Albright F (1942) Syndrome characterized by gynecomastia aspermatogenesis without A-Leydigism and increased excretion of follicle stimulating hormone. Journal of Clinical Endocrinology & Metabolism, 2, 615–627

Kumar P, Kumar N, Thakur DS, Patidar A (2010) Male hypogonadism: Symptoms and treatment. Journal of Advanced Pharmaceutical Technology & Research 1 (3):297

Lanciotti L, Cofini M, Leonardi A, Bertozzi M, Penta L, Esposito S (2019) Different Clinical Presentations and Management in Complete Androgen Insensitivity Syndrome (CAIS). International Journal of Environmental Research and Public Health 16 (7):1268

Liao J, Mehta L (2019) Molecular Genetics of Noonan Syndrome and RASopathies Pediatr Endocrinol Rev 16(Suppl 2):435-446

Mushannen T, Cortez P, Stanford FC, Singhal V (2019) Obesity and hypogonadism-a narrative review highlighting the need for high-quality data in adolescents. Children 6(5): 63

Niederberger C (2014) Re: Determinants of Testosterone Recovery after Bariatric Surgery: Is it Only a Matter of Reduction of Body Mass Index?. Journal of Urology 191 (3):753-754

Osta RE, Grandpre N, Monnin N, Hubert J, Koscinski I (2017) Hypogonadotropic hypogonadism in men with hereditary hemochromatosis. Basic and Clinical Andrology 27 :(1)

Prader-Willi Syndrome (2019) [(accessed on 12 December 2020)]; Available online: https://www.fpwr.org/about-prader-willi-syndrome#definitio

Rahnema CD, Lipshultz LI, Crosnoe LE, Kovac JR, Kim ED (2014) Anabolic steroid-induced hypogonadism: diagnosis and treatment. Fertility and Sterility 101: 1271-1279

Rao PM, Kelly, DM T. Jones H (2013) Testosterone and insulin resistance in the metabolic syndrome and T2DM in men. Nature Reviews Endocrinology 9 (8):479-493

Reed CE, Fenton SE (2013) Exposure to diethylstilbestrol during sensitive life stages: A legacy of heritable health effects. Birth Defects Research Part C: Embryo Today: Reviews 99 (2):134-146

Rodprasert W, Virtanen HE, Mäkelä JA, Toppari J (2019) Hypogonadism and Cryptorchidism. Front Endocrinol (Lausanne). 10: 906

Romina P. Grinspon, Nazareth Loreti, Débora Braslavsky, Clara Valeri, Helena Schteingart, María Gabriela Ballerini, Patricia Bedecarrás, Verónica Ambao, Silvia Gottlieb, María Gabriela Ropelato, Ignacio Bergadá, Stella M. Campo, Rodolfo A. Rey (2014) Spreading the Clinical Window for Diagnosing Fetal-Onset Hypogonadism in Boys. Frontiers in Endocrinology 5

Rubinstein AL, Carpenter DM (2017) Association Between Commonly Prescribed Opioids and Androgen Deficiency in Men: A Retrospective Cohort Analysis Pain Med 18(4): 637–644

Rubinstein AL, Carpenter DM, Minkoff JR (2013) Hypogonadism in Men With Chronic Pain Linked to the Use of Long-acting Rather Than Short-acting Opioids. The Clinical Clin J Pain 10:840-5

Smith HS, Elliott JA (2012) Opioid-induced androgen deficiency (OPIAD). Pain Physician (S3): ES145-156

Tajar A, Forti G, O'Neill TW, Lee DM, Silman AJ , Finn JD, Bartfai G, Boonen S, Casanueva FF, Giwercman A , Han TS , Kula K, Labrie F, Lean MEJ, Pendleton N, Punab M, Vanderschueren D, Huhtaniemi IT, F Wu FCW (2010) Characteristics of Secondary, Primary, and Compensated Hypogonadism in Aging Men: Evidence from the European Male Ageing Study. The Journal of Clinical Endocrinology & Metabolism 95 (4):1810-1818

Tsai EC, Boyko EJ, Leonetti DL, Fujimoto WY (2000) Low serum testosterone level as a predictor of increased visceral fat in Japanese-American men. International Journal of Obesity 24 (4):485-491

Vikas B, Maximilian M (2014) Noonan Syndrome. American Family Physician 89(1):37-43

Viswanathan V, Eugster EA (2011) Etiology and Treatment of Hypogonadism in Adolescents. Pediatric Clinics of North America 58 (5):1181-1200

Visootsak J, Graham JM (2006) Klinefelter syndrome and other sex chromosomal aneuploidies. Orphanet Journal of Rare Diseases 1: 42

Wong N, Levy M, Stephenson I (2017) Hypogonadism in the HIV-Infected Man Current Treatment Options in Infectious Diseases 9 , 104–116

Zhang Q, He H, Janjua MU, Wang F, Yang Y, Mo Z, Liu J, Jin P (2020) Identification of two novel mutations in three Chinese families with Kallmann syndrome using whole exome sequencing 2020; 52: (7) e13594

Effect of Obesity in the Cardiovascular System

6

Sabyasachi Das

Abstract

Overweight along with obesity are closely associated with an elevated risk of cardiovascular diseases as well as coronary artery disease. The subsequent consequence of obesity per se and the other associated medical conditions such as hypertension, glucose intolerance, diabetes, insulin resistance, sleep apnoea syndrome, etc. are put further deteriorating effect on human health. While, in the case of already existing cardiovascular diseases, the overweight and obese patients are often shows less mortality in compares to normal weight and underweight individual which is further termed as "obesity paradox." The underlying principle mechanism of the obesity paradox is not yet properly identified although several hypotheses are put forward for explaining the above-mentioned phenomena. Adipose tissue is now also well recognized as an endocrine organ, producing different potential chemical messengers, such as adipokines and adipocytokines. The knowledge of novel adipokines and adipocytokines are tremendously helped to understand the consequence of cardiovascular diseases (CVDs), obesity and metabolic syndrome. The precise roles played by this chemical messenger are particularly important to unfold the obesity-mediated consequence of inflammatory response, cardiac function, metabolism, satiety and other functions. Adipokines are particularly important to homoeostatic balance resulting in substantial benefits of angiogenesis, cardio-protection, endorsing endothelial function, and reducing hypertension, risk of atherosclerosis and inflammation. Considering the predominance of cardiovascular risk in obese individual, regular cardiology screening, and examination is of utmost important for the early diagnosis and treatment of sub-clinical medical conditions.

S. Das (✉)
Department of Physiology, Faculty of Medicine, Melaka Manipal Medical University College, Melaka, Malaysia

M. K. Kutty, A. Elengoe (eds.), *Obesity and its Impact on Health*,
https://doi.org/10.1007/978-981-33-6408-0_6

Keywords

Obesity · Cardiovascular disease · Heart failure · Hypertension · Obesity paradox · Adipokines

6.1 Introduction

Obesity is a serious health problem, growing significantly in the last couple of decades in developed as well as in developing countries throughout the world. The scourge of obesity has increased in epidemic proportions in children and adults. The body mass index (BMI) is solely used for evaluating the proportion of obesity. In adults, the term overweight is potentially defined as a body mass index (BMI) of 25–29.9 kg/m^2, on the contrary obesity is broadly defined as an accumulation of a surplus fat mass (FM) which impair normal health. Obesity is usually assessed through the determination of a BMI $\geq$30 kg/m^2, however, obesity is further subdivided into classes depending on the BMI (Class-I: BMI = 30.0–34.9; Class-II: BMI = 35.0–39.9, Class-III: BMI $\geq$40.0) to further classify the potential health risk (Carbone et al. 2019). Although BMI is extensively used to determine obesity but it gives no direct information on fat distribution, which is of higher significance in the assessment of different clinical conditions. Hence, other novel clinical indexes (measurements) have been introduced to predict the proportion of body fatness with more accuracy, i.e. waist circumference (WC), waist-to-hip ratio (WHR), and weight/height ratio. A waist circumference above 102 cm in men and above 88 cm in women are classified as the central obesity and involves with increased cardiovascular risk. More precisely, an estimation of waist/hip ratio above 0.9 in men and above 0.85 in women indicates central obesity (Csige et al. 2018). Following the Obesity Medicine Association, obesity defines as a chronic, relapsing, multifactorial, neuro-behavioral disease, where an increase in body fat endorses adipose tissue dysfunction and abnormal fat mass accumulation, resulting in adverse metabolic, biomechanical and psychosocial health consequences (Bays et al. 2018). Obesity has now designated as one of the critical problems in the United States, nearly 75% of adult populations are classified as obese or overweight compared with as low as 25% in the 1970s.

6.2 Millstone and Milestone Associated with Obesity

There are plenty of health millstone or adverse effects of obesity often observed as the excess fat mass characteristic ultimately increases the potential risk towards the most cardiovascular diseases (CVD), metabolic diseases like type-2 diabetes mellitus (DM), and their related major risks such as hypertension (HTN), heart failure (HF), and coronary heart disease (CHD) (Table 6.1) (Lavie et al. 2009). A recent report suggests the financial encumber in healthcare management and subsequent increase in morbidity and mortality resulting from complications of

Table 6.1 Obesity associated health hazards

A. On cardio vascular system	
1	Left ventricular hypertrophy
2	Concentric remodelling
3	Left atrial enlargement
B. Dyslipidemia	
1	Increased triglycerides
2	Increased low density lipid (LDL)
3	Increased very low density lipid (VLDL)
4	Increased apolipoprotein-B
5	Decreased high density lipid (HDL)
6	Decreased apolipoprotein-A1
7	Increased total cholesterol
C. Hypertension	
1	Increased blood volume
2	Altered peripheral resistance
D. Systolic and diastolic dysfunction	
E. Endothelial dysfunction	
F. Atrial fibrillation	
G. Coronary heart disease (CHD)	
H. Increased systemic inflammation	
I. Heart failure	
J. Metabolic disorders	
1.	Diabetes mellitus type-2
2	Glucose intolerance
K. Obstructive pulmonary disease	
L. Sleep apnoea	
M. Osteoarthritis	

obesity and CVD are increasing globally in last two decades producing an alarming sign toward control of the disease prognosis (Lavie et al. 2017). Recent advancement in cardiology research suggests, obesity itself a strong sovereign predictor of CHD as well as CVD even in the absence of other potential risk factors. Although obesity raises the risk of CVD in primary prevention, clinicians, as well as researchers, have historically assumed that surplus body mass would also be detrimental in secondary prevention settings. On the contrary, this assumption is not unavoidably correct as several retrospective and prospective reports subsequently demonstrated a potentially protective effect of obesity when it coexists with CVD; a phenomenon term as "obesity paradox." Recent investigations are trying to illuminate the role of obesity paradox in patients with heart failure (HF), CHD, CVD, hypertension, atrial fibrillation, and congenital heart disease.

6.3 Obesity and Heart Disease

There are three potential ways by which obesity contributes to heart disease. By an increase awareness on healthy life-style and taking good control over our health, we can combat not only with obesity but also against coronary heart disease. The potential ways are as follows:

1. Obesity can change the blood cholesterol level. Obesity has a great role toward increase in bad cholesterol (low-density lipid, LDL) as well as triglyceride level. On the contrary, it can reduce good cholesterol i.e. high-density lipid (HDL) which exerted an important role in removing LDL and thereby reduces the risk of coronary heart disease.
2. Obesity can cause a rise in blood pressure. An increase in bad cholesterol in the blood elevates the risk of lipid plaque formation in the inner wall of the blood vessel which leads to atherosclerosis and thereby elevation in blood pressure due to an increase in viscosity of blood and alteration in peripheral resistance within the blood vessel. Body mass index (BMI) is very high in obese person therefore, they need more blood to supply oxygen and nutrients to their bodies which leads to an elevation of blood volume and thereby increase in blood pressure. High blood pressure is one of the most common causes of a heart attack.
3. Obesity is one of the most leading causes of diabetes. Individuals with diabetes are found to be two to four times more affected by coronary heart disease. The metabolic dysfunction and inadequate filtration and absorption of different ions and metabolic end product in diabetes largely contribute toward the onset of heart diseases. Following American heart association data, at least 70 percent of individual aged above 60 or older with diabetes also has heart disease.

6.4 Obesity and Cardiovascular Risk: General Reflection

The potential risk for atherosclerotic vascular disease and thereby subsequent cardiovascular events are determined by several risk factors. The risk factors broadly are of two types, "endogenous" risk factors, including, genetic predisposition, gender, age, hormonal influences and blood pressure, whereas "exogenous" risk factors are modifiable factors like cigarette smoking, nutritional factors including dietary fat, body weight, physical activity, psychological stress and others (Gotto and Farmer 1992). Among these modifiable risk factors, obesity has become the major health concern accounting for its significant social and economic implications. Obesity is highly associated with increased mortality, including death from cardiovascular disease and heart failure. The advanced knowledge of global pandemic of obesity and its association with hypertension and CVD is of utmost importance to limit the mortality and morbidity associated with CVD and CHD.

6.5 Impact of Obesity on the Cardiovascular System

6.5.1 Effects of Obesity on Haemodynamics and Cardiovascular System

Numbers of adverse effects of obesity are reported on haemodynamics and cardiovascular structure of the heart (Fig. 6.1). By modulating the body's normal homeostasis, obesity substantially increases total blood volume through stimulating the bone marrow, and thereby increases the cardiac output, as well as cardiac workload. Usually, obese persons possess a much higher cardiac output but a significantly lesser level of total peripheral resistance at a given level of arterial pressure (Alpert 2001). The elevation of the cardiac output is caused by an increase of stroke volume, through elevated sympathetic activation, which also typically increases the heart rate to some extent (mildly). A leftward shift of the Frank-Starling curve is often observed in obese people because of increases in cardiac filling pressure and volume, thereby elevating cardiovascular workload. As a result of the above-mentioned phenomena, obese peoples are usually developed chronic hypertension than lean normal-weight persons. Typically gaining of weight is associated with an increase in arterial blood pressure. This substantial increase in filling pressure and volume in obese and overweight individuals usually develops dilation of the left ventricular chamber, which ultimately increase the risk of left ventricular hypertrophy (LVH), and other structural abnormalities of the heart, including concentric remodeling (CR) and concentric LVH, independent of individual age and arterial pressure. In addition to left ventricular structural anomalies, obesity also leads to left atrial enlargement, due to an increase in circulating blood volume and abnormal left ventricular diastolic filling. These abnormalities of the ventricle and atria potentially increase the risk of heart failure. Enlargement of left atria perhaps increases the risk

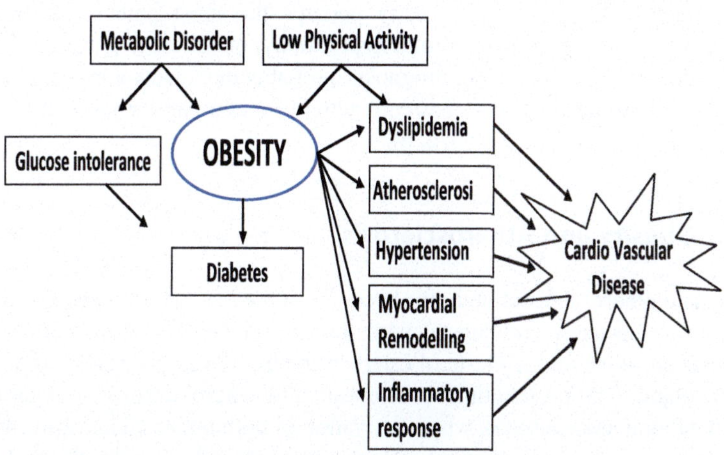

Fig. 6.1 The clinical consequences through which obesity increases the risk for cardiovascular disease

of atrial fibrillation (AF), whereas increasing left ventricular structural abnormalities elevated the risk of frequent and complex ventricular arrhythmias (Messerli et al. 1987; Wang et al. 2004). In addition, obesity also exerts numerous adverse effects on diastolic and systolic function (Chakko et al. 1991).

6.5.2 Obesity and Hypertension

In hypertension, the heart is pumping the blood with much greater force than normal. In general phenomenon, usually hypertension results in the thickening of ventricular walls without altering the ventricular chamber dilation, a process typically referred to as concentric remodeling or more specifically concentric left ventricular hypertrophy. On the contrary obese individuals are characterized by increase ventricular chamber dilation without a marked increase in the ventricular wall thickness, which typically leads to eccentric left ventricular hypertrophy (Lavie et al. 2007a, b).

Despite the greater prevalence of hypertension in obese individuals, recent reports have also suggested an obesity paradox (Uretsky et al. 2007). Some epidemiology-based studies showed lesser mortality in obese individuals who have suffering from CHD. In a follow-up study of 2 years with more than 22,500 patients, 30% less mortality was reported in obese and overweight patients, despite lesser effectual blood pressure control compared with the normal weight individual group. Despite hypertension lesser risk of stroke was reported in overweight patients. There is a significant "U-shaped" relationship between cardiovascular and non-cardiovascular mortality and BMI reported previously, which proves the excess mortality in both ex extremes of BMI i.e. very lean wight person and type-II category obese person (Stamler et al. 1991). Therefore the most accepted hypothesis referrers the association between BMI and major cardiovascular complications along with hypertension are results in a U-shaped curve. In addition, recent research in the cardiovascular system, suggest that although obesity perhaps the most potent risk factor for hypertension as well as left ventricular hypertrophy, but obese hypertensive patients may paradoxically possess better prognosis, probably due to lower systemic vascular resistance and an adequate plasma renin activity in comparison with leaner hypertensive patients (Lavie et al. 2007a, b).

6.5.3 Obesity and Atherosclerosis

Obesity and atherosclerosis both has been considered as lipid storage disorders for the past two decades, particularly with the accumulation of triglycerides in the peripheral fat tissue and cholesterol esters in the blood vessels as a form of atherosclerotic plaques. At present time, both obesity and atherosclerosis are regarded as a chronic inflammatory disease, where activation of both innate and adaptive immune processes are playing a crucial role (Rocha and Libby 2009). Obesity and atherosclerosis usually share the common pathophysiological factors including elevated free lipid droplets, particulates of oxidized LDL, fatty acids, and nitrous oxide which

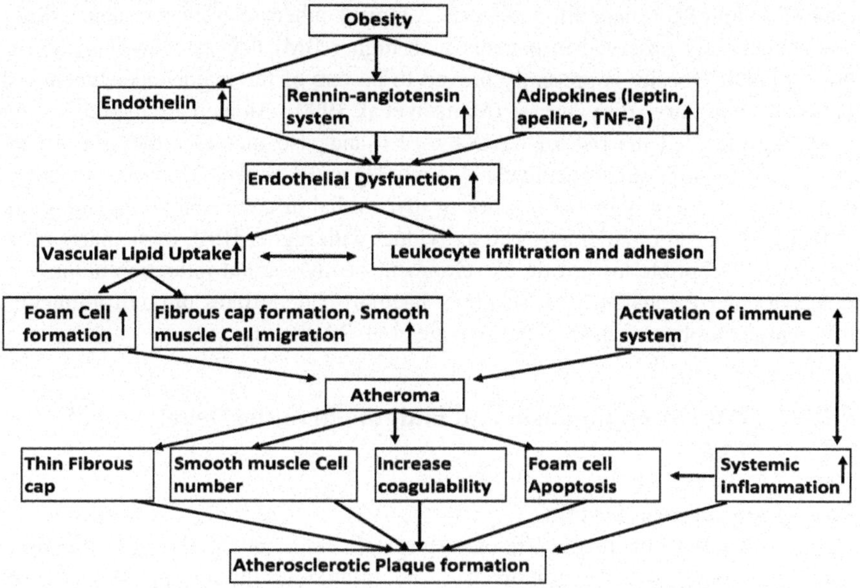

Fig. 6.2 The pathophysiology of atherosclerosis and coronary artery disease in obese individual. Role of endotheline and rennin-angiotensin system in obesity associated vasoconstriction, atherosclerosis and coronary artery disease

acts as an activator of the inflammatory process due to metabolic disorder and thereby triggers the disease pathogenesis. The persistent inflammatory response which subsequently triggers by obesity is one of the key factors towards the development of atherosclerosis, from the inception of early endothelial dysfunction to atherosclerotic plaques formation. Numerous adipokines (leptin, apeline, TNF-α, IL-6, and resistin) are constantly released from the fat tissue which subsequently induces endothelial dysfunction, reduced glucose tolerance, increase coagulability and systemic inflammation, thus facilitate the process of atherosclerosis (Shoelson et al. 2007). Obesity also elevates the level of C-reactive protein (CRP) which triggers the risk of myocardial infarction, peripheral vascular disease through increasing the level of anti-inflammatory cytokines (Ridker 2007) (Fig. 6.2). On the contrary, reduction of peripheral body fat induced the anti-inflammatory cytokines; elevate the release of adiponectin, chemerin, visfatin (insulin-sensitizing adipocytokine) whereas it significantly reduced the proinflammatory cytokines as well as insulin resistance.

6.5.4 Obesity and Coronary Artery Disease (CAD)

Coronary artery disease, especially coronary atherosclerosis is highly related to obesity. The inception of atherosclerosis happens several years before the manifestation of coronary artery disease. A typical, atherosclerotic vascular lesion is much frequently observed in patients with higher BMI than patients with normal body weight. Obesity is presently proved to be one of the major independent risk factors for coronary artery disease (Manson et al. 1990). An increase in body weight by 10 kg elevated the risk of CAD by 12% and also increases the systolic and diastolic blood pressure respectively by 5 mm Hg and 3 mmHg. Moreover, obesity is considered as the principal risk factor of myocardial infarction in young age-group individuals, ahead of smoking. The individual with higher BMI, particularly obese are more susceptible toward the development of myocardial infarction along with other vascular events such as ischemia, haemorrhagic strokes through mimicking different pathophysiological events (Csige et al. 2018) (Fig. 6.2).

6.5.4.1 Effect of Angiotensin and Endothelin in the Development of CAD

Role of Angiotensin System

Hyperactivation of the rennin-angiotensin system enhances the risk of cardiovascular disease. Subsequent inhibition of the renin-angiotensin system has efficiently delayed the development of different CAD including myocardial infarction. Activation of tissue renin-angiotensin it-self is one of the potential factors toward the inception and subsequent progression of abnormal cell growth, which leads to vascular hypertrophy, mesangial cell growth and thereby gradual development of atherosclerosis (Farmer and Torre-Amione 2001). Inhibitor of angiotensin-converting enzyme is now established as a cardiovascular protective agent through blocking angiotensin synthesis.

Endothelin

Endothelin is the major vasoconstrictor peptides synthesized from endothelial and other vascular cells. Apart from the regulation of vascular tone, endothelin-1 also acts as a potential growth stimulator and an activator of pro-inflammatory cytokines in different cells and tissue. Furthermore, endothelin-1 perhaps exerts potential effects in the onset of hypertension through activation of the renin-angiotensin system (Csige et al. 2018).

Effect of Obesity on the Rennin-Angiotensin System

Adipocytes, the functional unit of fat tissue contains the local rennin-angiotensin system as well as endothelin receptors. On the contrary, endothelin-1and angiotensin II enhances the synthesis of different adipokines (peptides that are secreted by adipocytes) such as leptin, resistin, etc. Rennin the key enzyme, converts angiotensin I from angiotensinogen, is found significantly higher in obese individual, suggesting that obesity subsequently activate the rennin-angiotensin system. Adipocytes stimulate the expression of genes encoding for different vasoactive proteins of the rennin-angiotensin system (Fig. 6.2).

Endothelin and Obesity

Greater expression of endothelin-1 is observed in obese individuals, at their cardiovascular system, including vasculature as well as in the cortex of kidney. In obese individuals, the elevation of endothelin-1 in plasma increases the permeability of the blood vessel which increases leukocyte adhesion and vascular lipid uptake. Substitution of lysine to asparagine at 198 codons of preproendothelin-1 gene induces vasoconstriction which results in obesity-induced hypertension in obese individual (Tiret et al. 1999). Furthermore, defective Endothelium-dependent vasomotion is also observed in obese individual. Endothelin-1 maintained the hyper-prostanoid-mediated vasoreactivity through the up-regulation of the thromboxane receptor gene in an obese individual (Traupe et al. 2002). The elevation of vasoconstriction is independent of systemic blood pressure, which subsequently proves obesity itself increases the vasoconstriction. Therefore, endotheline and rennin-angiotensin system directly contribute obesity-associated vasoconstriction, that perhaps further promote atherosclerosis, and hypertension in an obese individual (Fig. 6.2).

6.5.5 Obesity and Heart Failure

Heart failure is one of the major causes of death throughout the world. The incidence of heart failure and obesity are closely correlated. A one-unit rise of BMI elevated the risk of heart failure by 5% in men and 7% in female respectively. A higher BMI itself could be a high-risk factor of heart failure irrespective of age, sex and ethnicity. A recent report suggests 32%–49% of individuals suffering from heart failure are obese and 31–40% are overweight (Kenchaiah et al. 2002). Heart failure could develop 10 years earlier in overweight and obese individuals compared to normal BMI. The development of heart failure and the time duration of morbid obesity are also highly correlated. The prevalence of heart failure rises by 70% after 20 years of obesity whereas, after 30 years, the prevalence rate grows as high as 90%. In obese individuals, the subsequent structural and functional changes of heart alone contribute to deterioration in myocardial function, which is often referred to as "obesity cardiomyopathy" (Alpert et al. 2016).

Obesity may lead to heart failure through multiple direct and indirect ways. Higher BMI leads to the alteration of total haemodynamic of the body resulting in elevation of stroke volume, cardiac output as well as blood pressure. A rise of BMI by one unit involves an elevation of systolic blood pressure around 2 mmHg through the activation of the renin-angiotensin-aldosterone system and by increasing activity of the sympathetic nervous system (Ebong et al. 2014). Furthermore, obesity induced hypersecretion of aldosterone (through activation of angiotensin) as well as over expression of the mineralocorticoid receptors, ultimately initiates the interstitial cardiac-fibrosis, platelet aggregation, and endothelial dysfunction. On the contrary, obesity increase the blood volume which facilitates the venous backflow, enhances the ventricular preload causing an increase in ventricular wall tension and ultimately results in ventricular dilatation thereby developing left ventricular dysfunction. Prolonged hypertension on obese patients also increases the left ventricular

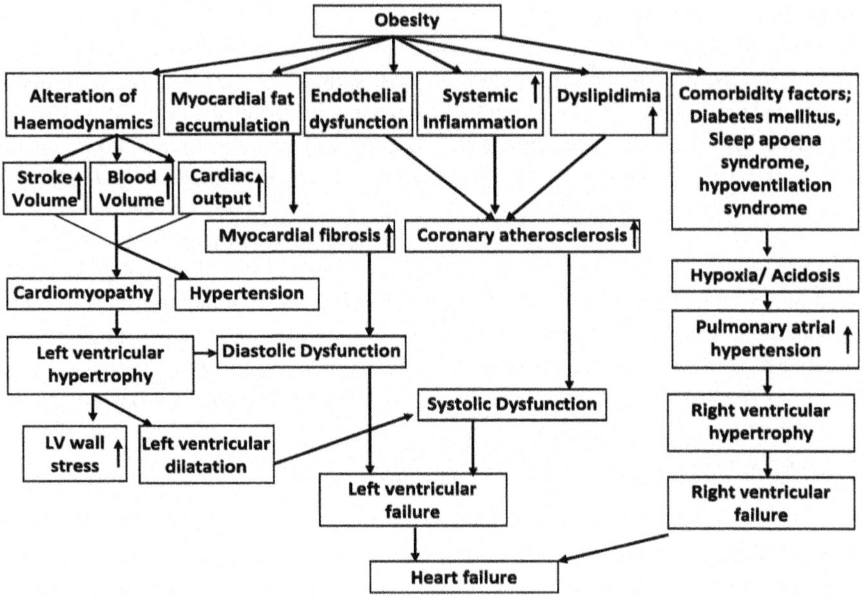

Fig. 6.3 The comprehensive pathophysiological mechanism of Heart failure in obesity. Obesity triggers multiple co-morbidity factors and other associated factors such as myocardial fat accumulation, endothelial dysfunction, coronary atherosclerosis, cardio myopathy, ventricular (both right and left) hypertrophy which finally leads to systolic and diastolic dysfunction followed by heart failure

afterload, which increases the risk of structural and electrical remodeling of myocardium. All these above-mentioned processes ultimately lead to the left ventricular hypertrophy and thereby diastolic and systolic ventricular dysfunction (Simone et al. 2002; Csige et al. 2018) (Fig. 6.3).

Inflammation exerts a huge role in heart failure. Obesity-induce the production of different Proinflammatory cytokines (TNF-α, IL-6, IL-8, IL-1, etc.), which played a pivotal role in the progression of heart failure. Different inflammatory mediators along with acute-phase proteins in coronary circulation cause the inception of myocardial fibrosis, which subsequently increases the myocardial stiffness and thereby leading to diastolic heart failure followed by and systolic heart failure (Torre-Amione et al. 1996). Proinflammatory cytokines also exert effects on metabolism, tissue structure, and release of different adipokines from Adipocytes (leptin, resistin, adiponectin, visfatin, etc.) which directly contributed to the myocardial transformation. In obese individuals, triglyceride accumulation in cardiac myocytes is commonly observed which eases the generation of toxic metabolites such as ceramide and diacylglycerol, thereby enhancing the apoptosis of cardiomyocytes (Karmazyn et al. 2008). Obesity-induced muscle atrophy and catabolism also plays a crucial role in the development of cardiovascular disease in an obese individual. Furthermore, obesity not only increases the chances of heart failure per se but also through modulating the associated co-morbidities. Reduce glucose tolerance, insulin resistance subsequently decrease the contractility of myocardium, while hyper-activation of obesity-induced renin-angiotensin-aldosterone system, can lead to hypertrophy as well as apoptosis of cardiac myocytes and thereby myocardial fibrosis. Therefore, obesity per se is an independent risk factor of coronary artery disease including different cardiac arrhythmias, which perhaps leads to the progression of heart failure (Fig. 6.3).

6.5.6 Obesity and Cardiac Arrhythmias

Several studies have demonstrated a close correlation between obesity and an increased risk of cardiac arrhythmias including sudden cardiac death. In fourth, Hippocrates concluded, "sudden death is more common in those who are naturally fat than in the lean."

6.5.6.1 Obesity and Stroke

Numerous reports suggested the close association between increasing BMI and stroke. The increase of BMI by one unit after overweight elevated the risk of ischemic stroke by 4% and haemorrhagic stroke by 6% (Kurth et al. 2002). This increased jeopardy related to stroke perhaps attributable to a greater prevalence of hypertension, a proinflammatory/prothrombotic state which attributes excess adipocytes accumulation, as well as increased atrial fibrillation.

6.5.6.2 Obesity and Atrial Fibrillation

Atrial fibrillation has contributed to the greatest medical significance among all cardiac arrhythmias. The global prevalence of atrial fibrillation is increasing rapidly as the burden of obesity and obesity-related disorders are subsequently increasing. Atrial fibrillation is affecting around 1–2% of the global adult population which is expected to increase by 2.5 folds in the next 20 years. Although the incidence of atrial fibrillation represents a strong correlation with age, pathophysiology of obesity and its adverse effect on haemodynamic, as well as unfavourable impact on left atria and left ventricle perhaps contribute toward the higher prevalence of Atrial fibrillation (Wanahita et al. 2008). Reports also suggests that obese individuals possess a 1.52 times greater risk toward the development of atrial fibrillation compared to normal BMI individuals (Tedrow et al. 2010; Wang et al. 2004). After the age of 65 years, the obese individuals possesses a 50% higher risk toward the development of atrial fibrillation. A unit raise of BMI after overweight elevates the frequency of newly developed atrial fibrillation by 4% which increases the risk of sudden cardiac death, stroke, and heart failure (Fig. 6.4).

6.5.6.3 Obesity and Ventricular Arrhythmias Including Sudden Cardiac Death

Numerous studies demonstrate the co-relationship between obesity and sudden cardiac death. Obesity per se is considered as an independent potential risk factor

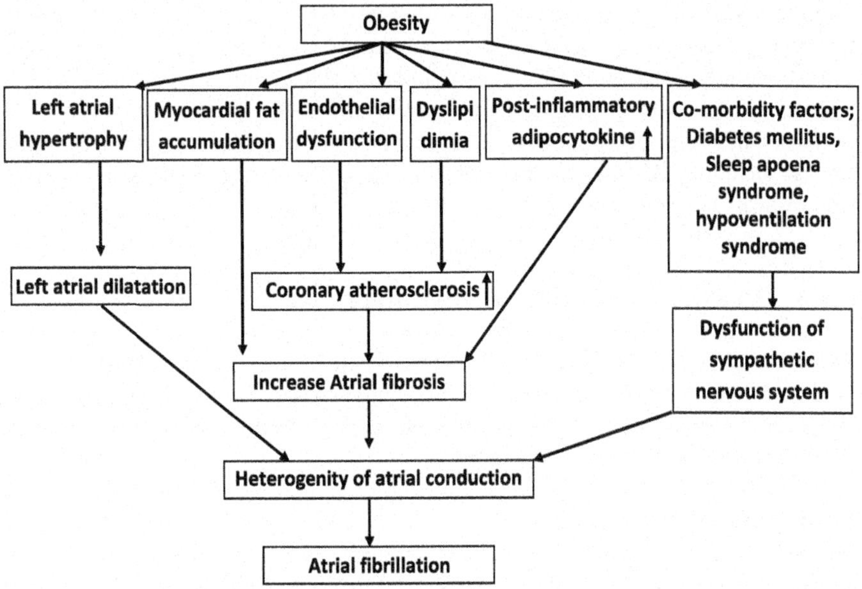

Fig. 6.4 The detailed pathophysiological mechanism of atrial fibrillation in obesity. Onset and subsequent development of atrial fibrillation depends on multifactorial phenomena. Prolonged left atrial hypertrophy followed by increased atrial fibrosis and sympathetic nervous system dysfunction finally leads to heterogenity of atrial conduction and thereby atrial fibrillation

in the progression of ventricular tachyarrhythmias. The structural and electrical remodeling of ventricular myocardium in obese patients results toward the development of left ventricular hypertrophy and followed by systolic and diastolic ventricular dysfunctions. Elevation of epicardial fat, myocardial hypertrophy and fibrosis are the footsteps of this pathological process. Progressive heart failure is one of the commonest causes of death in obesity cardiomyopathy. An increase of electrical irritability in the obese individual is probably leading toward the onset of frequent and complex ventricular dysrhythmias, even without any left ventricular dysfunction or heart failure (Lavie et al. 2009). Obesity is perhaps associated with prolonged and inhomogeneous ventricular repolarization, which can subsequently develops prolongation of QT interval and QT interval corrected to the heart rate (QTc). A positive correlation between BMI and corrected QT (QTc) interval is established previously. A prolonged QTc is usually used to predict the increased mortality even in healthy populations. An increase in prolonged QTc is usually observed in patients with severe grade-2 obesity (el-Gamal et al. 1995). Obesity markedly increases the number of abnormal late potentials. The elevation of late potentials is subsequently linked with the increased threat of sudden cardiac death. The inception of subsequent late potentials perhaps related to some pathophysiological alterations, such as obesity-induced cardiomyopathy, which altered the function of voltage-dependent potassium channels, and autonomic dysregulation.

6.5.6.4 Obesity and Structural and Electrical Remodeling of Myocardium

Prolonged overweight and obesity are results in several structural and functional alterations that plays an important role in arrhythmogenesis. The most common consequences of obesity are left atrial dilation and dysfunction. An increase of 5 mm diameter in the left atrium gives raises the chance of paroxysmal atrial fibrillation by 1.4 times. Furthermore, recent reports suggested that epicardial fat contributes to the development of atrial interstitial fibrosis through its paracrine effect. The elevations of epicardial fat, infiltration of the myocardium with adipocytes, as well as atrial fibrosis are together result in heterogeneous atrial pulse conduction, which contributes to endo- and epicardial electrical dissociation (Maesen et al. 2013). These processes assist in the development of atrial re-entry, which further serves as the electrophysiological basis of atrial fibrillation. Elevation of body fatty is strongly correlated with myocardial hypoxia (Fig. 6.4).

Atrial fibrillation in the obese individuals shortens the refractory period of pacemaker cells present in the atrium. Excess fat tissue triggers the alterations of ECG pattern, especially abnormal ECG pattern with higher amplitude P waves, elevation of the PR interval, alteration of P wave terminal force, etc. Similar electrical changes of heart are usually observed in atrial interstitial fibrosis, congestive heart failure, hypertension, and myocardial infarction.

6.5.7 Obesity and Sleep Apnoea

Obesity is one of the principal causes of alveolar hypoventilation as well as obstructive sleep apnea. Obstructive sleep apnoea perhaps contributes toward the pathogenesis of hypertension, dysrhythmias, pulmonary hypertension, myocardial infarction, stroke and increased inflammatory response but the details underlying mechanisms are yet to be elucidate.

6.5.8 Obesity and Venous Disease

In obesity, lack of physical activity, elevation of intravascular volume as well as a greater volume of lymphatic overload usually leads to the venous insufficiency and oedema.

6.6 The Obesity Paradox

Although it is now well-established fact that obesity enhances the risk of cardiovascular diseases but for those individuals who have already developed the disease, excess fat and obesity exerts a favourable prognosis for them. This particular phenomenon has been observed in several cardiovascular diseases' including acute and chronic heart failures, myocardial infarction, atrial fibrillation is known as the obesity paradox (Csige et al. 2018).

6.6.1 The Obesity Paradox and Heart Failure

It is undeniable fact that obesity exerts a pivotal risk factor for heart failure as well as heart failure-related diseases, but after the occurrence of heart failure, obesity also exerts some protective effects. Such a paradoxical association is especially observed in patients with class-I obesity. After heart failure, the obesity paradox is primarily observed in patients with advanced disease condition which coexists with overweight or obesity. Overweight and obese patients exert improved prognosis compared to normal BMI and underweight patients.

Following a meta-analysis involving 28,209 patients with heart failure cases, during 2.7 years of the follow-up period, overweight patients' shows, 16% lower overall mortality and 19% lower cardiovascular mortality compared to normal BMI patients. On the contrary, obese patients show even more favourable data as 33% lower overall mortality was observed along with 40% lower cardiovascular mortality compared to normal BMI patients. Recent reports have also confirmed the obesity paradox in heart failure with preserved left ventricular ejection flow as well as with heart failure with reduced left ventricular ejection flow. The underlying molecular mechanism by which obesity improves the prognosis in the individual with heart failure is yet to be understood. Although, several hypotheses have been proposed (e.g. an elevation of lean mass in obese individuals) that could plays a pivotal role toward improved long-term outcomes. After the occurrence of heart failure, excess lean mass in an obese individual probably allows greater cardio-respiratory fitness which is closely related to improved prognosis after heart failure. Recent evidence confirmed that peak oxygen consumption (VO_2) is more sturdily dependent on lean mass not on the body mass. The underlying precise mechanisms of cardio-respiratory fitness and the precise role of lean mass during the prognosis are yet to be understood (Carbone et al. 2019) (Fig. 6.5).

6.6.2 Obesity Paradox in CHD

Obesity is a key risk factor for the progression of coronary heart disease (CHD), but patients with already established CHD with higher BMI are represented a more favourable prognosis, establishing an obesity paradox. The precise protective mechanism of coronary heart disease in an obese individual is not clearly understood. Numerous epidemiologic reports suggest that the alteration in body composition in obese individual perhaps mediate some prognostic benefits. Similarly like the obesity paradox in heart failure, the elevation of lean mass in obese individuals likely to exert a protective effect. In addition to the elevation of lean mass, the surplus adiposity perhaps providing protective action against already established CHD thereby exerting a more favourable prognosis compared to normal BMI and

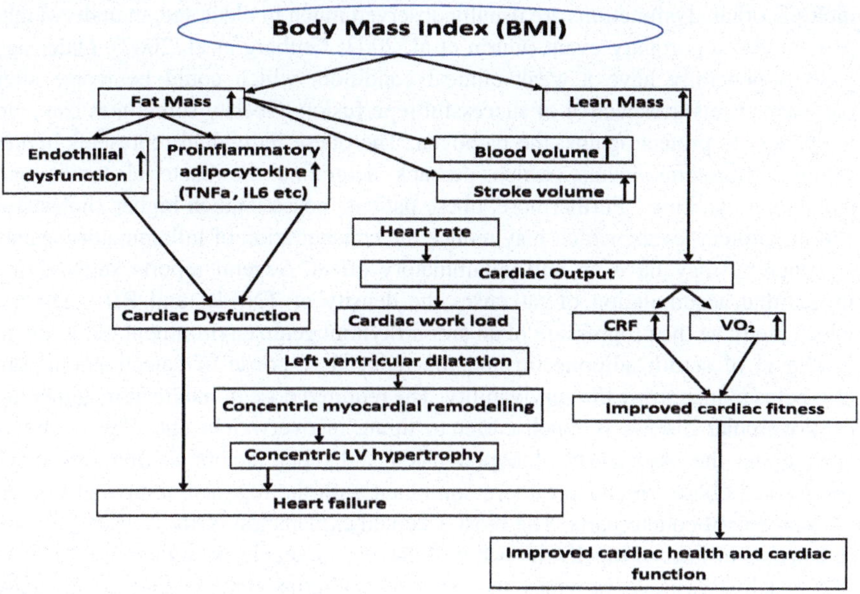

Fig. 6.5 The pathophysiology of obesity paradox

underweight individuals. Elevation of adiposity is likely to be more beneficial in individuals with low systemic inflammation. The precise way by which low systemic inflammation increases in the obese individual is unknown to us till date. Adipose tissue per se can induce the production of pro-inflammatory cytokines and thereby elevates the low systemic inflammation. Therefore, it is speculated that precise therapeutics targeting of systemic inflammation may induce a much better beneficial effect in overweight and obese individuals (Carbone et al. 2019). Furthermore, like heart failure cases, higher cardiorespiratory fitness is another most potent predictor for better clinical outcomes in established CHD patients, independent of BMI or total fat mass or visceral adiposity.

6.6.3 Obesity Paradox: A Critical Explanation

Obesity and overweight are well-known phenomena of metabolic syndrome, which leads to enhanced cardiovascular risk, myocardial infarction, endothelial dysfunction, heart failure, inflammation, and atherosclerosis. Therefore, the most critical question of importance itself is "what can be the possible explanation behind the better prognosis of already established CVD and CHD cases in overweight and obese patients compared to normal BMI or underweight patients?" Cardiovascular diseases are generally diagnosed and treated much earlier in overweight and obese patients than in normal or underweight patients. Several reports suggest different comorbid conditions such as malignant diseases, heart failure, malnutrition, or multiple organ dysfunctions are usually observed much in older age, in underweight patient after a coronary event (Gurm et al. 2002; Gruberg et al. 2002). Older age patients generally have a weak clinical condition, which could be worse after coronary events irrespective of successful reperfusion. Obesity sometimes can help the person to protect against the pathogen. The obese individual represents higher serum lipoproteins levels, which perhaps neutralize the bacterial toxins and circulating cytokines. Furthermore, obese patients possess much higher cholesterol and lipoprotein levels, which may reduce the concentration of inflammatory agents and thereby may have an anti-inflammatory effect. Several reports suggest that myocardial accumulation of fat raises the density of TNF-RI and RII receptors, which facilitate the progression of an anti-arrhythmogenic environment. Subsequent low level of serum adiponectin and the reduced response of catecholamine are also helps in increasing the survivability. The required dose of medication in treating cardiovascular disease is much easier to titrate in overweight and obese patients considering the association of hypertension and other factors. Particularly overweight and obese patients are more compliant with the regimen compared to their normal-weight counterparts. The most accepted explanation of the obesity paradox refers that in such critically ill patients; the fat molecule which mobilized from excess adipose tissue provides the required additional energy and significantly prevents the lean tissue from wasting more efficiently than exogenous nutrients (Goossens et al. 2017). A metabolic cardiac remodeling event usually occurs after heart failure, which results in impaired fatty acid oxidation and rapid enhancement of glucose uptake and glycolysis. An individual with heart failure cases shows variation in mitochondrial dynamics, respiratory capacity, and lack of ATP synthesis which ultimately plays an important role in chronic cardiac energy deficit. Enhanced fatty acid utilization in the obese individual through dietary modification significantly ameliorates mitochondrial fragmentation and cardiac dysfunction. On the contrary, antithrombotic medications are usually used in standard control doses instead of subsequent adjusted to body weight, therefore there is a high risk of overdosing in underweight thin patients, which may increase the chances of bleeding complications and thereby contribute to higher mortality. Recent reports of the obesity paradox, explaining that obese individual possesses larger blood vessels which is advantageous in the course of primary coronary intervention, while worse results are obtained in underweight or normal patients with narrowed blood vessels (Schunkert et al. 1999) (Fig. 6.5).

On the contrary, if we consider the outcomes of cardiopulmonary stress, the so-called "favourable effect" of higher BMI is mostly disappearing. Peak oxygen consumption (VO_2) and good cardiorespiratory fitness are the most positive predictor of prolonger survivability among the patients with heart failure. The obesity paradox of BMI is no longer been applicable in diabetic patients with heart failure; similarly, the protective role of BMI is not valid in relation to Peak oxygen consumption (VO_2). Even though an obesity paradox exists, which favours more toward the prognosis of already established cardiovascular disease in overweight and obese patients than leaner patients, but the huge constellation of data originated from cardiovascular disease research still supports the purposeful weight reduction for the prevention and treatment of cardiovascular diseases (Fig. 6.5).

6.7 Endocrine Function of Adipose Tissue

Obese individual contains a very big quantity of peripheral and central adipose tissue. Adipose tissue is now also well recognized as an endocrine organ, producing different potential chemical messengers, such as adipokines and adipocytokines. The knowledge of novel adipokines and adipocytokines are tremendously helped us to understand the consequence of cardiovascular diseases (CVDs), obesity and metabolic syndrome. A total of 30–50% of adipose tissue is comprised of adipocytes. Adipokines (leptin, adiponectin, visfatin, etc.) are the group of polypeptides secreted by adipocytes while adipocytokines {macrophage inflammatory protein (MIP), MCP-1, TNFα, IL6, IL8, etc.} refers to the polypeptides released from the stroma vascular cells. Interestingly, adipokine apelin and resistin may secret form both adipocytes and stroma vascular cells. As an endocrine organ, white adipose tissue exerts a plethora of body functions including cardiovascular function, glucose and lipid metabolism, thermogenesis, energy and feeding regulation, neuroendocrine function, reproduction, and immunity (Mattu and Randeva 2013).

6.8 Obesity and Adipokine Function

Obesity is principally characterised by a perpetual, low-grade systemic proinflammatory state causing subsequent hypertrophy and hyperplasia of fat cells which ultimately leads to an imbalance of adipokines secretion. Adipokines exert their actions through different signalling pathways and chemical mediators. Their individual effects are achieved successively through the regulation of different mediators along with the crosstalk with other adipokines.

6.8.1 Adiponectin

Adiponectin is primarily synthesised by adipocytes but it is also secreted from skeletal muscle and cardiomyocytes. Adiponectin is belonging to the collagen superfamily protein having a molecular weight of 30 kDa with a length of 244 amino acid. 0.01% of total plasma proteins are Adiponectin. Adiponectin intercedes its biological actions through three different receptors: Adipo-R1 (ubiquitously expressed in skeletal muscle), Adipo-R2 (mostly in the liver) and T-cadherin. All three adiponectin receptors are primarily expressed in cardiac tissue (Mattu and Randeva 2013).

6.8.1.1 Adiponectin and Obesity

Plasma adiponectin levels are significantly reduced in obese individuals than the non-obese individual. Therefore, a negative correlation is found between body mass index (BMI) and plasma adiponectin level. Furthermore, plasma adiponectin concentrations have also been negatively correlated with percent body fat and intra-abdominal fat. Plasma adiponectin levels are gradually increased with weight loss and with insulin-sensitising drugs, therefore it appearing to be inversely proportional to the insulin resistance and obesity. The reason for reduced adiponectin in obese individual remains unclear. Suppression at the transcription level or decreased secretion of specific pro-inflammatory cytokines may reduce the adiponectin level. In obese individual pro-inflammatory cytokines IL-6 is up-regulated and perhaps exerts some negative effects at transcriptional level thereby reduced the expression adiponectin mRNA from 3 T3-L1 adipocytes in the experimental model (Hopkins et al. 2007).

6.8.1.2 Adiponectin and Cardiovascular Disease

The role of adiponectin in the progression of cardiovascular disease is not clearly understood, although some studies show that a low level of plasma adiponectin is closely associated with cardiovascular disease including acute myocardial infarction. Some of other studies also shows high plasma adiponectin level is probably associated with myocardial infarction and coronary heart disease in diabetic males. Most of the studies suggested that prolonged hypo-adiponectinemia is associated with the inception of cardiovascular diseases which is prevalent in obese individuals. The plasma levels of adiponectin are inversely correlated with different cardiovascular risk factors, including hypertension, atherosclerosis and hyperlipidemia (Ouchi et al. 2003; Iwashima et al. 2004). The precise role of plasma adiponectin in the inception of coronary artery disease is controversial, as most of the study is not able to found a strong correlation, although it appears that low levels of adiponectin are not much reliable marker for this disease. Hypertrophic cardiomyopathy is commonly associated with obesity. Reduced plasma adiponectin is closely associated with the inception of left ventricular hypertrophy in obese patients.

The epidemiological data of plasma adiponectin and its relation with the progression of heart failure is particularly complicated mostly due to the obesity paradox. High plasma adiponectin levels are observed in patients who are died from heart failure (Hopkins et al. 2007). Most probably a paradoxical relationship may exist, as low plasma adiponectin, favours survival, at end-stage heart failure in obese patients. Therefore, future detailed studies are very much important to enlighten the detailed action of adiponectin on cardiovascular diseases (Fig. 6.6).

6.8.1.3 Adiponectin and Atherosclerosis

Recent experimental reports suggest that adiponectin plays a crucial role in the progression of obesity-related vascular diseases including atherosclerosis. During the inception of atherosclerosis, circulating monocytes adhere to activated endothelial cells; thereby monocytes differentiate into macrophages which leads to accretion

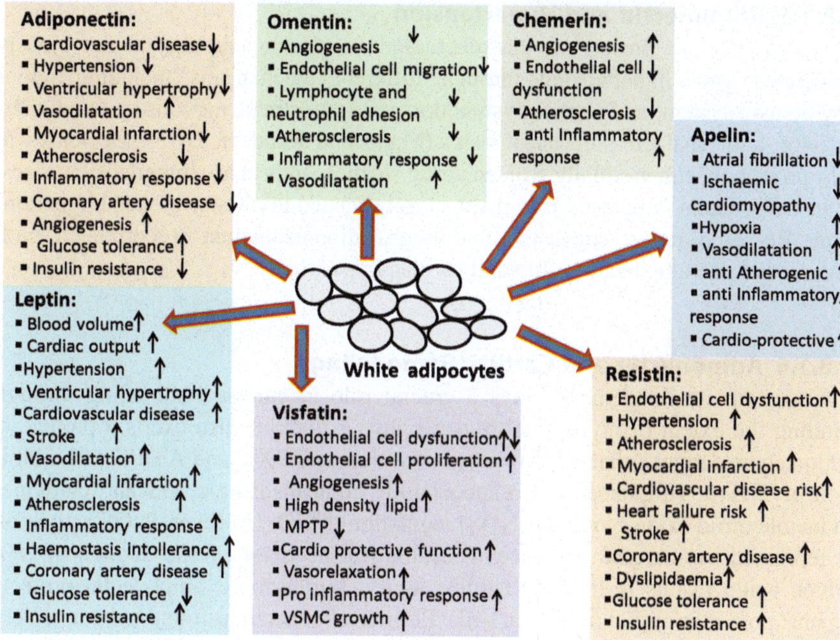

Fig. 6.6 Different physiological function (especially on cardio vascular system) of Adipokines

of LDL-rich foam cells. Adiponectin precisely regulates several key steps to limit the atherogenic process.

Adiponectin particularly inhibits the NF-κB in endothelial cells thereby slows the development of atherosclerosis. Plasma adiponectin further subsequently inhibits the transformation of macrophages into foam cells and thereby reduces the oxidation of LDL and suppresses the expression of class-A scavenger receptors in human macrophages. Furthermore, adiponectin reduces TNF-α production and increases the expression of anti-inflammatory cytokine IL-10 in macrophages. Following the above methods, suppression of NF-kB halts TNF-α induced process and thereby reduced atherosclerosis.

6.8.1.4 Adiponectin and Endothelial Function

Vascular endothelium per se is an important paracrine organ in regulating the vascular tone, smooth muscle cell proliferation as well as inflammation. Generation of NO is extremely critical as well as crucial to execute the functions of vasoconstriction, vasodilatation, leukocyte transmigration, smooth muscle proliferation in normal physiological conditions. Adiponectin particularly controls the key enzyme decisive for the production of NO-eNOS, through activation of phospho-inositol 3 kinase (PI3K) and AMPK which reduces vascular smooth muscle cell proliferation and other endothelial functions.

6.8.1.5 Adiponectin and Hypertension

It is proved with precession that decreased in plasma adiponectin are closely associated with hypertension through different mechanisms including renin-angiotensin system, endothelial dysfunction and sympathetic nervous system hyper-activity, and renal pressure natriuresis (Mattu and Randeva 2013). Decreases in plasma adiponectin gradually worsen aortic stiffness in primary hypertension cases. Thus adiponectin deficiency is perhaps closely related to obesity-induced hypertension. Research report suggested that hypo adiponectinemia is correlated with myocardial damage, heart failure and cardiac hypertrophy.

6.8.1.6 Adiponectin and Cardiac Remodeling

Plasma adiponectin perhaps plays a crucial role in cardiac remodeling through limiting the extent of myocardial hypertrophy. Adipokine also exerts a protective action against myocardial infarction by activation of COX-2 and AMPK and inhibition of TNF-α. Adiponectin also successfully controlled the eNOS and cytokine-inducible nitric oxide synthase (iNOS) regulation of NO. However, iNOS activation is repressed by increase adiponectin resulting in decrease overproduction of NO, which could results in increase cardiac injury. Furthermore, adiponectin protects against post-myocardial infarction (MI). Plasma adiponectin is found to be inversely related to insulin resistance and the severity of coronary artery disease.

Therefore, adiponectin is termed as a 'good' adipokine because of its cardioprotective effects, anti-atherogenic, anti-inflammatory, anti-diabetic and promotion of good endothelial function. It also reduced the risk of coronary artery disease along with post-MI protection and promotes angiogenesis.

6.8.2 Leptin and Its Receptors

Leptin is a 16-kDa hormone adipokine, mainly secreted by white adepocytes in a diurnal pulsatile manner (increase secretion in early morning and evening.), principally acts on hypothalamus usually decrease appetite and increase energy expenditure. Leptin exerts a variety of action particularly in energy balance and metabolism, cardiovascular homeostasis, hypertension, atherosclerosis, and coronary heart disease, vascular inflammation, oxidative stress, and type 2 diabetes mellitus. Leptin is acts through its trans-membrane receptors (LEPR, or OB-R). Six different isoforms of LEPR or OB-R are (LEPRa, b, c, d, e, and f, also known as OB-Ra, b, c, d, e, and f) observed with their distinct biological function. All of OB-R isoforms represents nearly similar structures with 840 amino acids long highly glycosylated extracellular domain which comprises six subdomains. Leptin hormone contains multiple binding sites, especially of cytokine receptor homologous domains-II (CRHII) and immuno globulin D (IgD) to interact with LEPR, followed by LEPR activation to regulate numerous biological functions.

6.8.2.1 Leptin and Obesity

The level of plasma leptin is directly correlated with total volume of adipocytes or body fat mass. After 4–6 h of heavy meal serum leptin level is increases as an effect of insulin stimulated glucose utilization which subsequently up-regulated the leptin mRNA transcription and thereafter secretion of leptin from white adipose tissue. Enhanced adipocytes volume in obese individual, results in increase of plasma leptin volume which ultimately triggers hyperinsulinemia and elevated turn-over of serum cortisol. Although some other factors like elevated TNF-α, oestrogen and free fatty-acid can also stimulate the secretion of leptin in a lesser to moderate extent. On the contrary, increase in serum catecholamine level as well as lower serum glucose and insulin are results in reduction of serum leptin concentration (Fig. 6.6).

6.8.2.2 Leptin Resistance

Most of the obese individuals are hyperleptinaemic. Instead of high serum leptin concentration the energy expenditure and utilization are usually reduced significantly in obese individual, which termed as Leptin resistance. Leptin resistance is significantly linked with hypothalamic inflammation, insulin resistance, and metabolic disturbances which subsequently elevated the risk toward inception of hypertension, cardiovascular diseases and metabolic syndromes (Wannamethee et al. 2007). In leptin resistance, neurones of hypothalamus became less sensitive to leptin despite of large quantity of serum leptin. The underlying mechanisms of leptin resistance are yet to be elucidated. Some of the serum protein is directly modulating the leptin activity. Recent reports suggests, human C-reactive protein (CRP) inhibits the binding of leptin receptors to leptin and subsequently blocks the energy expenditure processes. Moreover, obesity induced leptin resistance are mainly associated with different genetic factors, especially due to genetic mutations in leptin receptors (LEPR), less mRNA expression of leptin receptors, impairment of second messenger associated with LEPR thereby altered the function of leptin. Reduce transport of serum leptin across blood-brain barrier in obese individual could be one of the potential cause of leptin resistance. Leptin receptor mutations in obese individual are mostly autosomal recessive trait. Numerous leptin receptor mutations are identified at the extracellular domain, more specifically at the leptin-binding site or at its activation domain. Interestingly, not a single mutation is observed at intracellular domain of leptin receptor. Substitution of glycine to alanine at codon 2597 of exon 16 is the most common LEPR mutation observed in severe obesity, which exactly results in, the premature termination of translation. Functional genomics analysis reveals, there are another 4 LEPR mutation which are strongly correlated with Leptin resistance and subsequently characterized with early-onset of obesity, hyperphagia, altered immune-function, and delayed pubertal growth and development. The heterozygous R612H mutation in LEPR modulates the activation of second messenger STAT3 in response to leptin binding. On the contrary single nucleotide polymorphism at codon A409E, W664R, and H684P results complete loss of leptin induced cell signalling. Furthermore obesity induced leptin receptor mutations is also strongly associated with insulin resistance and type-2 diabetes mellitus. Interestingly, long term elevation of serum leptin increases the probability to develop diet induced obesity, while increase in while adipocytes further increases the serum leptin level and thereby aggravating the leptin resistance, which usually called as leptin induced leptin resistance (Poetsch et al. 2020).

6.9 Conclusion

In this book chapter, the author looks into the impact of obesity on cardiovascular system. A heart-related disease such as heart attack affects young people nowadays. Therefore, young people should take healthy diet food and regular exercises Considering the predominance of cardiovascular risk in obese individual, regular cardiology screening, and examination is of utmost important for the early diagnosis and treatment of sub-clinical medical conditions.

References

Alpert MA, Omran J, Bostick BP (2016) Effects of obesity on cardiovascular hemodynamics, cardiac morphology, and ventricular function. Curr Obes Rep 5(4): 424-434

Alpert MA (2001) Obesity cardiomyopathy: pathophysiology and evolution of the clinical syndrome. Am J Med Sci 321: 225-236

Bays HE, Seger J, Primack C (2018) Obesity algorithm, presented by the Obesity Medicine Association. obesityalgorithm.org; Accessed on 10th June 2018

Carbone S, Canada JM, Billingsley HE, Siddiqui MS, Elagizi A, Lavie CJ (2019) Obesity paradox in cardiovascular disease: where do we stand? Vasc Health Risk Manag 15: 89-100

Chakko S, Mayor M, Allison MD, Kessler KM, Materson BJ, Myerburg RJ (1991) Abnormal left ventricular diastolic filling in eccentric left ventricular hypertrophy of obesity. Am J Cardiol 68: 95-98

Csige I, Ujvárosy D, Szab Z, Lőrincz I, Paragh G, Harangi M, Somodi S (2018) The impact of obesity on the cardiovascular system. J Diabetes Res 1-12

Ebong IA, Goff Jr DC, Rodriguez CJ, Chen H, Bertoni AG (2014) Mechanisms of heart failure in obesity. Obes Res Clin Pract 8(6) e540-e548

El-Gamal A, Gallagher D, Nawras A (1995) Effects of obesity on QT, RR, and QTc intervals. Am J Cardiol 75: 956-959

Goossens C, Marques MB, Derde S (2017) Premorbid obesity, but not nutrition, prevents critical illness-induced muscle wasting and weakness. J Cachexia, Sarcopenia and Muscle 8(1): 89-101

Gotto AM, Farmer JA (1992) Risk factors for coronary artery disease. In E. Brauuwald (Ed.), Heart disease. A textbook of cardiovascular medicine (pp. 1125–1160) (2nd ed.) Philadelphia: W.B. Saunders Co

Gruberg L, Weissman NJ, Waksman R (2002) The impact of obesity on the short-term and long-term outcomes after percutaneous coronary intervention: the obesity paradox? J Amer Coll Cardiol 39(4) 578-84

Gurm HS, Whitlow PL, Kip KE (2002) The impact of body mass index on short- and long-term outcomes in patients undergoing coronary revascularization: insights from the bypass angioplasty revascularization investigation (BARI). J Am Coll Cardiol 39(5) 834-840

Farmer JA, Torre-Amione G (2001) The renin–angiotensin system as a risk factor for coronary artery disease. Curr Atheroscler Rep 3: 117-124

Hopkins TA, Ouchi N, Shibata R, Walsh K (2007) Adiponectin actions in the cardiovascular system. Cardiovas Res 74: 11-18

Iwashima Y, Katsuya T, Ishikawa K, Ouchi N, Ohishi M, Sugimoto K (2004) Hypoadiponectinemia is an independent risk factor for hypertension. Hypertension 43:1318-1323

Karmazyn M, Purdham DM, Rajapurohitam V, Zeidan A (2008) Signalling mechanisms underlying the metabolic and other effects of adipokines on the heart. Cardiovasc Res 79(2): 279-286

Kenchaiah S, Evans JC, Levy D (2002) Obesity and the risk of heart failure. N Engl J Med 347(5): 305-313

Kurth T, Gaziano JM, Berger K (2002) Body mass index and the risk of stroke in men. Arch Intern Med 162: 2557-2562

Lavie CJ, Milani RV, Ventura HO (2007a) Obesity, heart disease, and favorable prognosis-truth or paradox? Am J Med 120: 825- 826

Lavie CJ, Milani RV, Ventura HO, Cardenas GA, Mehra MR, Messerli FH (2007b) Disparate effects of left ventricular geometry and obesity on mortality in patients with preserved left ventricular ejection fraction. Am J Cardiol 100: 1460-1464

Lavie CJ, Milani RV, Ventura HO (2009) Obesity and cardiovascular disease risk factor, paradox, and impact of weight loss. J Am Coll Cardiol 53(21): 1925-1932

Lavie CJ, Arena R, Alpert MA, Milani RV, Ventura HO (2017) Management of cardiovascular diseases in patients with obesity. Nat Rev Cardiol 15(1): 45-56

Maesen B, Zeemering S, Afonso C (2013) Rearrangement of atrial bundle architecture and consequent changes in anisotropy of conduction constitute the 3-dimensional substrate for atrial fibrillation. Circ Arrhythm Electrophysiol 6(5): 967-975

Manson JAE, Colditz GA, Stampfer MJ (1990) A prospective study of obesity and risk of coronary heart disease in women. N Engl J Med 322(13): 882-889

Mattu HS, Randeva HS (2013) Role of adipokines in cardiovascular disease. Journal of Endocrinology. 216: T17-T36

Messerli FH, Nunez BD, Ventura HO, Snyder DW (1987) Overweight and sudden death: increased ventricular ectopy in cardiomyopathy of obesity. Arch Intern 147: 1725-1728

Ouchi N, Kihara S, Funahashi T, Matsuzawa Y, Walsh K (2003) Obesity, adiponectin and vascular inflammatory disease. Curr Opin Lipidol 14: 561-566

Poetsch MS, Strano A, Guan K (2020) Role of leptin in cardiovascular disease. Front Endocrinol 11:1–13

Ridker PM (2007) C-reactive protein and the prediction of cardiovascular events among those at intermediate risk: moving an inflammatory hypothesis toward consensus. J Am Coll Cardiol 49 (21): 2129-2138

Rocha VZ, Libby P (2009) Obesity, inflammation, and atherosclerosis. Nat Rev Cardiol 6(6): 399-409

Schunkert H, Harrell L, Palacios IF (1999) Implications of small reference vessel diameter in patients undergoing percutaneous coronary revascularization. J Am Coll Cardiol 34(1): 40-48

Shoelson SE, Herrero L, Naaz A (2007) Obesity, inflammation, and insulin resistance. Gastroenterology 132(6): 2169-2180

Simone G, de Palmieri V, Bella JN (2002) Association of left ventricular hypertrophy with metabolic risk factors: the Hyper GEN study. J Hypertens 20(2): 323-331

Stamler R, Ford CE, Stamler J (1991) Why do lean hypertensive patients have higher mortality rates than other hypertensive patients? Findings of the hypertension detection and follow-up program. Hypertension 17: 553- 564

Tiret L, Poirier O, Hallet V, McDonagh T, Morrison C, McMurray J, Dargie H, Arveiler D, Ruidavets J, Luc G, Evans A, Cambien F (1999) The Lys198Asn polymorphism in the endothelin-1 gene is associated with blood pressure in overweight people. Hypertension 33: 1169-1174

Torre-Amione G, Kapadia G, Benedict C, Oral H,Young JB, Mann DL (1996) Proinflammatory cytokine levels in patients with depressed left ventricular ejection fraction: a report from the studies of left ventricular dysfunction (SOLVD). J Am Coll Cardiol 27(5): 1201-1206

Tedrow WB, Conen D, Ridker PM. (2010) The long-and short-term impact of elevated body mass index on the risk of new atrial fibrillation: the WHS (women's health study). J Am Coll Cardiol 55(21): 2319-2327

Traupe T, Lang M, Goettsch W, Muenter K, Morawietz H, Vetter W, Barton M (2002) Obesity increases prostanoid-mediated vasoconstriction and vascular thromboxane receptor gene expression. J Hypertens 11: 2239-2245

Uretsky S, Messerli FH, Bangalore S (2007) Obesity paradox in patients with hypertension and coronary artery disease. Am J Med 120: 863-870

Wanahita N, Messerli FH, Bangalore S (2008) Atrial fibrillation and obesity-results of a meta-analysis. Am Heart J 155: 310-355

Wang TJ, Parise H, Levy D (2004) Obesity and the risk of new-onset atrial fibrillation. JAMA 292 (20): 2471-2477

Wannamethee SG, Tchernova J, Whincup P, Lowe GD, Kelly A, Rumley A (2007) Plasma leptin: associations with metabolic, inflammatory and haemostatic risk factors for cardiovascular disease. Atherosclerosis 191:418–426, 443

Impact of Obesity on Vaccination

7

Abdullah Yahya Al-Mahdi and Asita Elengoe

Abstract

Worldwide, obesity is a serious health problem. It contributes to chronic diseases such as cardiovascular diseases (e.g. heart attack, heart-related diseases), diabetes mellitus (DM) type 2, respiratory diseases, reproductive disorders (e.g. polycystic ovarian syndrome (PCOS), endometritis), kidney-related diseases, osteoarthritis, etc. However, obesity also is associated with an immune response. Obese people have low immune-competence. Numerous studies on human subjects and animal models were carried out to determine the impact of obesity on the immune response. They found that the production of antibodies (eg. IgG) and functional activity of natural killer (NK) cells in obese people is low. In this book chapter, we will discuss the impact of obesity on vaccination. Although the development of effective vaccines to protect against numerous preventable infectious diseases such as Hepatitis A, Hepatitis B, Hepatitis C, rabies, influenza, tetanus, etc. are successful; but the obese population has been found to respond poorly to vaccines.

Keywords

Obesity · Immune response · Vaccination

A. Y. Al-Mahdi
Department of Microbiology, Faculty of Medicine, Lincoln University College, Selangor, Malaysia

A. Elengoe (✉)
Biotechnology Department, Faculty of Science, Lincoln University College, Petaling Jaya, Selangor, Malaysia

M. K. Kutty, A. Elengoe (eds.), *Obesity and its Impact on Health*,
https://doi.org/10.1007/978-981-33-6408-0_7

7.1 Introduction

Since the subject of obesity and its global incidence along with the comorbidities has been described in the introduction I feel it is redundant here for any further discussion (WHO consultation on obesity 2000). While obesity-related insulin resistance (diabetes mellitus (DM) type 2) coupled with chronic inflammation is the major factor in the pathogenesis, of the metabolic syndrome, it is associated with decreased immune-competence and poor antibody response to hepatitis B plasma vaccine (Marti et al. 2001).

Obesity, as defined elsewhere, denotes the storage of surplus energy in the white adipose tissue (WAT). The regulatory cells in adipose tissue; suppresses the activity of the lymphocytes (B-lymphocytes and T-lymphocytes such as T-cytotoxic, T-helper); macrophages; and natural killer (NK) cells (Ouchi et al. 2011; Han and Levings 2013). T-helper lymphocytes are actively involved in the production of cytokines that induce the activity of natural killer cells, macrophages and B lymphocytes. While T-cytotoxic lymphocytes play an important role in the killing of abnormal or infected cells (Janeway et al. 2005). Moreover, metabolic inflammation in adipose tissue interferes with the mechanism of insulin production by beta islets of Langerhans and causes the resistance of insulin. This develops diabetes mellitus (DM) type 2. Grade of obesity is closely linked with the poor immune response (Marti et al. 2001; Cristiane et al. 2009).

7.2 The Impact of Obesity on Vaccination

Although the development of effective vaccines to protect against numerous infectious diseases such as Hepatitis A, Hepatitis B, Hepatitis C, rabies, influenza, etc. are successful; but some populations have been found to respond poorly to vaccines available now. This will escalates the risk for vaccine-preventable diseases such as Polio, Tuberculosis, human papillomavirus (HPV) vaccine, tetanus, etc. Some researchers revealed that obesity affects vaccine immunogenicity and efficacy. The clinical trials proved that obesity is one of the factors to develop a poor vaccine-induced immune response in humans (Table 7.1). Obesity happens via the excess accumulation of lipids into white adipose tissue (WAT) via adipocytes production. This process is called a body-mass index (BMI) ≥ 30 kg/m^2. Obese people (children, adult and elderly people) have a poor immune system. They have low production of antibody (IgG) and nature killer (NK) cells. The understanding of the mechanisms of the poor vaccine-induced immune response in obese people should be studied so that obese people might be protected against preventable diseases.

Table 7.1 Human subject studies of immune responses to vaccines

Author	Year	Vaccine	No. of subjects	Study outcome
Weber et al.	1985	Hepatitis B	194	Inadequate of antibody response to the hepatitis B virus plasma vaccine
Eliakim et al.	2006	Tetanus	15	Reduced anti-tetanus IgG antibodies
Banga et al.	2014	Rabies	603	Lack of antibody titer to the rabies virus plasma vaccine
Napolitano et al.	2009	Influenza A (H1N1)	10	Lowered anti-influenza A IgG antibodies
Hanslik et al.	2010	Influenza A (H1N1)	1266	Reduced anti-influenza A IgG antibodies
Louie et al.	2011	Influenza A (H1N1)	228	Decreased anti-influenza A IgG antibodies
Peralta et al.	2010	Influenza A (H1N1)	652	Lack of antibody titer to the influenza A virus plasma vaccine
Morgan et al.	2010	Influenza A (H1N1)	361	Decreased anti-influenza A IgG antibodies
Diaz et al.	2011	Influenza A (H1N1)	144	Reduced anti-influenza A IgG antibodies
Kwong et al.	2011	Influenza A (H1N1)	82,545	Decreased anti-influenza A IgG antibodies
Lin et al.	2007	Hepatitis B	5406	Inadequate of antibody response to the hepatitis B virus plasma vaccine
Jack et al.	1999	Hepatitis B	700	Lack of antibody response to the hepatitis B virus plasma vaccine
Young et al.	2013	Hepatitis B	123	Low level of antibody response to the hepatitis B virus plasma vaccine
Van der Wielen et al.	2006	Hepatitis B	501	Reduction in antibody response to the hepatitis B virus plasma vaccine
Reuman et al.	1997	Hepatitis A	100	Inadequate of antibody response to the hepatitis A virus plasma vaccine
Coleman et al.	2013	Influenza A (H1N1)	2623	Inadequate of antibody response to the influenza A virus plasma vaccine
Yang et al.	2013	Influenza A (H1N1)	4048	Inadequate of antibody response to the influenza A virus plasma vaccine
Callahan et al.	2014	Influenza A (H1N1)	178	Decreased anti-influenza A IgG antibodies
Talbot et al.	2012	Influenza A (H1N1)	415	Decreased anti-influenza A IgG antibodies
Pondo et al.	2014	Anthrax	1563	Decreased anthrax IgG antibodies
Paich et al.	2013	Influenza A (H1N1)	111	Decreased anti-influenza A IgG antibodies
Segerstrom et al.	2012	Influenza A (H1N1)	134	Lowered anti-influenza A IgG antibodies
Sheridan et al.	2012	Influenza A (H1N1)	499	Reduced anti-influenza A IgG antibodies

(continued)

Table 7.1 (continued)

Author	Year	Vaccine	No. of subjects	Study outcome
Averhoff et al.	1998	Hepatitis B	2293	Decreased hepatitis B virus plasma vaccine
Wood et al.	1993	Hepatitis B	595	Low level of hepatitis B virus plasma vaccine

7.3 Conclusion

Obesity is a major public health problem. We cannot ignore the suboptimal vaccine-induced immune responses found in obese people. While we continue to find solutions to fight against obesity via strict and healthy food diet and regular exercises, obese people remain at higher risk for vaccine-preventable illnesses. The inadequate research data regarding the impact of obesity on vaccine-induced immunity is alarming. There is a need to understand how the mechanism of the immune cell works in obese people. This understanding will aid in the design of an effective vaccine for obese people. More research on vaccine immunogenicity and efficacy in obese people should be carried out. Human subjects (infants, young children, adults and elderly people) and animal models (rodents) might be included in the research studies. There is always a need for data to develop improved treatments for obese people.

References

Averhoff F, Mahoney F, Coleman P, Schatz G, Hurwitz E, Margolis H (1998) Immunogenicity of Hepatitis B vaccines: implications for persons at occupational risk of Hepatitis B virus infection. American Journal of Preventive Medicine 15: 1-8

Banga N, Guss P, Banga A, Rosenman KD (2014) Incidence and variables associated with inadequate antibody titers after pre-exposure rabies vaccination among veterinary medical students. Vaccine 32(8): 979-983

Callahan ST, Wolff M, Hill HR, Edwards KM (2014) Impact of body mass index on immunogenicity of pandemic H1N1 Vaccine in children and adults. The Journal of Infectious Diseases 210 (8): 1270-1274

Coleman LA, Waring SC, Irving SA, Vandermause M, Shay DK, Belongia EA (2013) Evaluation of obesity as an independent risk factor for medically attended laboratory-confirmed influenza. Influenza and other Respiratory Viruses 7(2): 160-167

Cristiane MM, Ivo M, Jean Pierre SP, Luiz VR, and Alfredo H (2009) Impact of adiposity on immunological parameters. Arquivos Brasileiros de Endocrinologia & Metabologia 53(2): 183-189

Díaz E, Rodríguez A, Martin-Loeches I, Lorente L, Del Mar Martín M, Pozo JC, Montejo JC, Estella A, Arenzana Á, Rello J (2011) Impact of obesity in patients infected with 2009 influenza A (H1N1). Chest 139(2): 382-386

Eliakim A, Schwindt C, Zaldivar F, Casali P, Cooper DM (2006) Reduced tetanus antibody titers in overweight children. Autoimmunity 39(2): 137-141

Han JM, and Levings MK (2013) Immune regulation in obesity-associated adipose inflammation. Journal of Immunology 191(2): 527-532

Hanslik T, Boelle PY, Flahault A (2010) Preliminary estimation of risk factors for admission to intensive care units and for death in patients infected with A (H1N1)2009 influenza virus, France, 2009-2010. PLoS Currents 2: RRN1150

Jack D, Hall AJ, Maine N, Mendy M, Whittle HC (1999) What level of hepatitis B antibody is protective? The Journal of Infectious Diseases 179(2): 489-492

Janeway CA, Travers P, Walport M, and Sholmchik MJ (2005) Immunobiology: the immune system in health and disease. 6th edition. Garland Science Publishing Inc.: New York

Lin YC, Hsiao ST, Chen JD (2007) Sonographic fatty liver and hepatitis B virus carrier status: synergistic effect on liver damage in Taiwanese adults. World Journal of Gastroenterology 13 (12): 1805-1810

Louie JK, Acosta M, Samuel MC, Schechter R, Vugia DJ, Harriman K, Matyas BT (2011) A novel risk factor for a novel virus: obesity and 2009 pandemic influenza A (H1N1). Clinical Infectious Diseases 52(3): 301-312

Marti A, Marcos A, and Martinez JA (2001) Obesity and immune function relationships. Obesity Reviews 2(2): 131-140

Morgan OW, Bramley A, Fowlkes A, Freedman DS, Taylor TH, Gargiullo P, Belay B, Jain S, Cox C, Kamimoto L, Fiore A, Finelli L, Olsen SJ, Fry AM (2010) Morbid obesity as a risk factor for hospitalization and death due to 2009 pandemic influenza A(H1N1) disease. PLoS ONE 5(3): e9694

Napolitano LM, Park PK, Sihler KC, Papadimos T, Chenoweth C, Cinti S, Zalewski C, Sharangpani R, Somsel P, Wells E, Fry AM (2009) Intensive-care patients with severe novel influenza A (H1N1) virus infection-Michigan, June 2009. Morbidity and Mortality Weekly Report 58(27): 749-752

Ouchi N, Parker JL, Lugus JJ, and Walsh K (2011) Adipokines in inflammation and metabolic disease. Nature Reviews Immunology 11(2): 85-97

Paich HA, Sheridan PA, Handy J, Karlsson EA, Schultz-Cherry S, Hudgens MG (2013) Overweight and obese adult humans have a defective cellular immune response to pandemic H1N1 influenza A virus. Obesity (Silver Spring) 21(11): 2377-2386

Peralta PS, Cortes-García M, Vicente-Herrero M, Castrillo-Villamandos C, Arias-Bohigas P, Pachon-del Amo I, Sierra-Moros M J (2010) Risk factors for disease severity among hospitalised patients with 2009 pandemic influenza A (H1N1) in Spain, April-December 2009. Europe Surveillance 15(38): 19667

Pondo T, Rose CE, Jr, Martin SW, Keitel WA, Keyserling HL, Babcock J (2014) Evaluation of sex, race, body mass index and pre-vaccination serum progesterone levels and post-vaccination serum anti-anthrax protective immunoglobulin G on injection site adverse events following anthrax vaccine adsorbed (AVA) in the CDC AVA human clinical trial. Vaccine 32(28): 3548-3554

Reuman PD, Kubilis P, Hurni W, Brown L, Nalin D (1997) The effect of age and weight on the response to formalin inactivated, alum-adjuvanted hepatitis A vaccine in healthy adults. Vaccine 15(10): 1157-1161

Segerstrom SC, Hardy JK, Evans DR, Greenberg RN (2012) Vulnerability, distress, and immune response to vaccination in older adults. Brain Behaviour Immunology 26(5): 747-753

Sheridan PA, Paich HA, Handy J, Karlsson EA, Hudgens MG, Sammon AB (2012) Obesity is associated with impaired immune response to influenza vaccination in humans. International Journal of Obesity 36(8): 1072-1077

Talbot HK, Coleman LA, Crimin K, Zhu Y, Rock MT, Meece J (2012) Association between obesity and vulnerability and serologic response to influenza vaccination in older adults. Vaccine 30 (26): 3937-3343

Van der Wielen M, Van Damme P, Chlibek R, Smetana J, von Sonnenburg F (2006) Hepatitis A/B vaccination of adults over 40 years old: comparison of three vaccine regimens and effect of influencing factors. Vaccine 24(26): 5509-5515

Yang WH, Dionne M, Kyle M, Aggarwal N, Li P, Madariaga M (2013) Long-term immunogenicity of an AS03-adjuvanted influenza A(H1N1)pdm09 vaccine in young and elderly adults: an observer-blind, randomized trial. Vaccine 31(40): 4389-4397

Young KM, Gray CM, Bekker L-G (2013) Is obesity a risk factor for vaccine non-responsiveness? PLoS ONE 8(12): e82779

Weber DJ, Rutala WA, Samsa GP, Santimaw JE, Lemon SM (1985) Obesity as a predictor of poor antibody response to hepatitis B plasma vaccine. JAMA 254(22): 3187-3189

WHO consultation on obesity (2000) Obesity: preventing and managing the global epidemic. WHO Tech Rep Series no 894. Geneva, Switzerland: World Health Organization

Wood RC, MacDonald KL, White KE, Hedberg CW, Hanson M, Osterholm MT (1993) Risk factors for lack of detectable antibody following hepatitis B vaccination of Minnesota health care workers. JAMA 270: 2935-2939

Obesity-Linked Diseases (Comorbidities) 8

Al-Abed Ali Ahmed Al-Abed

Abstract

Due to healthy lifestyle changes, obesity is one of the serious health problems and has a high prevalence worldwide. In the last three decades, obesity was increased rapidly as it is linked with many comorbidities leading to escalate the risk of disability, illness and fatality, especially by non-communicable diseases. Excessive weight gain is obviously due to overconsumption of calories and these extra calories are stored as fat tissue which may lead to develop certain diseases such as cardiovascular diseases (CVD) and hypertension. Obesity has an important role and impact on cardiovascular diseases (CVD), such as coronary heart disease (CHD), heart failure (HF), and atrial fibrillation. Obesity as an essential cause of hypertension and linked with diabetes mellitus, the combination of obesity with anyone of these diseases leads to develop most of the manifestations of CVD. It was reported that increasing body weight will increases the risk of type 2 diabetes at all ages as the body of overweight or obese people make more effort and inability to control blood sugar levels by using insulin and have a high chance to develop diabetes. In addition, abdominal obesity was reported to be related to 2–3 times escalate the risk of hypertension and can be declined with a reduction of waist size. It was also found that there is an association between the presence the risks of dementia and Alzheimer's disease (AD) with obesity. Moreover, obesity has a relationship with rheumatoid arthritis (RA), dyslipidemia, reproductive disorders, respiratory diseases, renal diseases, sleep disorders and many other diseases.

Keywords

Obesity · Diseases · Comorbidities

A.-A. A. A. Al-Abed (✉)
Community Medicine Department, Faculty of Medicine, Lincoln University College, Petaling Jaya, Malaysia

M. K. Kutty, A. Elengoe (eds.), *Obesity and its Impact on Health*,
https://doi.org/10.1007/978-981-33-6408-0_8

8.1 Introduction

Obesity is a serious health problem and is becoming pandemic. The prevalence of obesity globally has approximately increased twice in the last three decades and is associated with escalating risks of disability, illness and fatality as reported by World Health Organization (WHO) (Branca 2010). Obesity is classified as one of the top ten risks among other non-communicable diseases due to healthy lifestyle changes (Subhija et al. 2012). The most common non-communicable diseases such as cardiovascular diseases, diabetes and hypertension, are the main problems associated with obesity. Currently, millions of people including different ages and both genders are obese. It was reported that more than 1 billion adults and nearly 10% of children have been categorized as overweight or obese (Haslam and James 2005). In the United States only, around 70% of adults were overweight or obese (Luma and Anwar 2011). Needless to emphasize that obesity impacts quality of life and serious health issues that are potentially life-threatening, including arteriosclerosis, heart disease, certain cancers, dyslipidemia, hypertension (HTN), depression, metabolic syndrome, sleep disorders, physical and psychological disturbance. Thus, these complications will augment the risk of developing type II diabetes mellitus, dyspepsia (Indigestion), Intestinal disorders, Impotence and inferiority, short- and long-term health problems for pregnant mother and child and other health issues (Carl et al. 2009; Shu-Zhong et al. 2016). This chapter attempts to focus on and discuss some of the obesity-related diseases. It is felt necessary to briefly deal with the causes of obesity.

8.2 Causes of Obesity

Obesity is a multifactorial chronic inflammatory disease, the causes of which are, in short, the imbalance between energy consumption and energy depletion, which further exacerbates this imbalance, the complex interactions between unhealthy eating habits, sedentary lifestyle and genetic factors. Obesity is a multifactorial chronic disease that may occur due to the unbalancing between energy consumption and energy production. Compounding this imbalance may form a complex interaction between unhealthy dietary habits, sedentary lifestyle and genetic factors (Galbete et al. 2013). It is almost impossible to define and determine the specific cause of obesity, but it is related largely to excessive food consumption. Dealing with obese patients has demonstrated that most patients share specific habits. All of which are characterized by common traits;

1. an addiction to excessive food consumption.
2. Improper Manner of Ingesting Food
3. Sedentary Lifestyle
4. Consumption of High-Calorie Food or High Sugar Intake
5. Psychological Factors Such as Depression, Low Esteem
6. Genetic Factors

7. Family History

The global incidence of obesity, a modern scourge, is escalating exponentially over the last two decades. People from different high-socioeconomic groups in both developing and developed countries are now victims of the disease. Moreover, socioeconomic conditions coupled with affluence and technological advances have rendered people to become victims of sedentary life. Obesity perse is a problem of great magnitude and this is becoming perilous because of its comorbidities which have been described below.

8.3 Obesity-Related Diseases (Comorbid Conditions)

To comprehend the obesity-linked problems, it is imperative to define the distinction between overweight and obesity. Excess weight is not synonymous with obesity and every fat individual has excess weight. However, overweight signifies a few extra pounds over the body's needs. For example, if a person of average height has 5–10 pounds of excess weight, he is not obese. On the other hand, weight gain in an obese person is almost constant. A person is labeled as obese if he has an exceptional surplus accumulation of fat increasing his weight. Excessive weight exerts pressure on all organs of the obese individuals continuously, and consequently, he or she becomes exposed to several health risks. Obesity obviously, is the dismal consequence of an imbalance between energy consumption and energy disbursement (Rosenbaum and Leibel 1998). Genetic susceptibility is one of the contributing factors of gaining weight; however, contradictory in some reports have denied it, due to eating habits and physical activity may have a more substantial role in the quantity of weight gained (Racette et al. 2005; Nestle and Jacobson 2000)

Excessive weight gain is obviously due to overconsumption of calories as compared to those required for the body, and these extra calories are stored as fat tissue (Rosenbaum and Leibel 1998). A fat-rich diet has a greater impact on weight gain compared to the excess of calories. Most of the calories that are stored in adipose tissue increase body weight; (Life 2014) as well as the cholesterol level in circulation (Kwok et al. 2010). The excess fat in circulation predisposes to the development of atherosclerosis (Ledwozyw et al. 1986) of medium and large arteries (Matsuzawa 2005) which may be responsible for certain diseases such as cardiovascular diseases and hypertension. Although the effects of anti-obesity drugs become more developed nowadays, the effect of obesity on some other related non-communicable diseases and metabolic disorders is most likely to become more significant in the future and escalating the prevalence by time (John et al. 2015). The most important obesity related-comorbidities will be discussed in this chapter as below.

8.3.1 Cardiovascular Diseases

Obesity has an important role and impact on cardiovascular diseases (CVD), such as coronary heart disease (CHD), heart failure (HF), and atrial fibrillation. Needless to reiterate that the risk of heart disease is associated with increased LDL-cholesterol levels and decreased HDL-cholesterol levels, which are responsible for the development of coronary atherosclerosis leading to myocardial infarction. The cardiac workload becomes greater among obese patients which raises the total blood volume and cardiac and reduces the total peripheral resistance (Alpert 2001; Messerli 1982). In obesity, the cardiac output augmented with increased stroke volume; increase sympathetic activation causes a typically mild increase in heart rate (Messerli 1987). Also, the risk of diabetes is augmented among obese by diminishing glucose tolerance and susceptible to develop of left ventricular hypertrophy. Obesity-related left ventricular hypertrophy cause elevating the load of pumping the blood throughout the body by the heart. It is estimated that each pound of fat may need a mile of extra blood vessels to supply nutrients and oxygen (Ostlund et al. 1990; Lauer et al. 1991).

Obesity is known to be an essential cause of hypertension, and the combination of obesity and hypertension is acknowledged as a paramount hazard to develop most of the manifestations of CVD (Lewis et al. 2013; Thomas et al. 2005; Luma and Anwar 2011). In this context, it is important to refer to the phenomenon of "obesity paradox" reported by many studies that exhibit a negative correlation between overweight and obese people with developing of cardiovascular diseases, including heart failure, and coronary heart disease, hypertensive heart disease and peripheral arterial disease (Carl et al. 2009; Alpert 2001).

A recent study demonstrated that overweight or obese adults aged 40–59 years are exposed to the ominous risk of developing the cardiovascular disease with a shortened lifespan as compared with their normal-weight peers. Even younger overweight persons are predisposed to develop cardiovascular disease (Sadiya et al. 2018). A plethora of studies have emphasized the relation of obesity with cardiovascular disease (Freedman et al. 2002; Skyler and Oddo 2002; Gregg et al. 2005) as obesity is associated with some subclinical vascular diseases with markers of subclinical CVD (Lauer et al. 1991; Oren et al. 2003; Kuller et al. 1994).

8.3.1.1 Heart Failure (HF)

Obesity is related to structural and functional changes in the heart. Therefore the prevalence and incidence of heart failure were increased. It was reported that heart failure has more prevalent among obese and overweight individuals compared to normal-weight people (Carl et al. 2009). In addition, prevalence increment of heart failure was reported due to the long duration of morbid obesity (Alpert et al. 1997). In a study of 5881, participants were followed up for 14-years and found that by increasing 1 unit of BMI, the risk of developing Heart Failure (HF) elevated by 5% among men and 7% among women (Kenchaiah et al. 2002). HF among obese may be occurred due to an elevation of exposure to other HF risk factors such as coronary artery disease (CAD), hypertension, diabetes mellitus, metabolic syndrome and some other conditions. Undefinable, myocardial dysfunction can be developed among obese people who are considered to have "obesity cardiomyopathy" (Ashrafian et al. 2008).

8.3.1.2 Atrial Fibrillation (AF) or Flutter

Currently, the prevalence of atrial fibrillation (AF) or flutter is mounting and is projected to be increased by 2.5-fold in 2050. Sixteen studies comprising an aggregate of 123,000 patients were assessed for the impact of obesity; the result of studies revealed that the occurrence of AF among obese patients had approximately 50% elevated the risk of AF development that correlated with elevating BMI. However, studying of 44,647 patients with post-cardiac surgery did not confirm this observation (Wanahita et al. 2008). It was also reported that aging with the improved prognosis of HTN, CHD, and HF, and obesity, may contribute to augment the risk of AF (Wang et al. 2004).

Atrial fibrillation (AF) or flutter was developed among 372 men and 181 among women. By increasing 1 unit of BMI, the adjusted hazard ratio for AF or flutter was 1.08 among men and 1.06 among women while this ratio among overweight was 1.75 in men and 1.39 in women if normal-weight group was a reference. It was found that escalating the risk of atrial fibrillation or flutter was related to overweight and obesity (Lars et al. 2005).

8.3.1.3 Stroke

Subjects with stroke or those who developed ischemic heart disease have a high ratio of waist to hip circumference compared to normal controls (Larsson et al. 1984). It was reported that BMI associated with stroke as increasing 1-unit of BMI, there was an increase of 4% in the risk of ischemic stroke and 6% for hemorrhagic stroke. This augmented hazard of a stroke may be referred to a higher prevalence of HTN, a prothrombotic/proinflammatory state that accompanies surplus adipose tissue accumulation and increasing of AF as well (Poirier et al. 2006). This subject has been discussed in detail in another chapter.

8.3.2 Diabetes Mellitus

Diabetes signifies the failure of the body to produce any or adequate insulin to control hyperglycemia. The pancreas loses progressively the ability to produce insulin and even if insulin is present, the individual suffers from diabetes due to insulin resistance. Type 1 and type 2 diabetes are the main types of diabetes mellitus, the latter being commoner than the former; type 2 diabetes mellitus constitutes approximately 90–95% of all diabetes cases (Center for Disease Control and Prevention 2019). The current term "diabesity" connotes obesity dependent diabetes.

One of the most significant risk factors linked with type 2 diabetes is obesity. It is estimated by body mass index (BMI), which is calculated by weight and height. The waist circumference is a good indicator of non-insulin-dependent diabetes mellitus (NIDDM) compared to BMI, waist/hip ratio over other anthropometric measurements (Wei et al. 1997). BMI has a robust link to diabetes and insulin resistance. A strong positive association has been noted between overall obesity for those with BMI ≥ 35 kg/m^2 and the hazard of diabetes compared with men with BMI <23.0 kg/m^2 (June et al. 1994). Studies have demonstrated an association between obesity with pancreatic β-cell dysfunction and insulin resistance which controls blood glucose level (Kasuga 2006).

Obesity intensifies the risk of diabetes due to diminishing glucose tolerance. In overweight or obese people, their bodies make more effort and inability to control blood sugar levels by using insulin and have a high chance to develop diabetes. Escalating body weight increases the risk of type 2 diabetes at all ages. It was reported that nearly 90% of people living with type 2-diabetes are overweight or obese. Overweight individuals are more prone to develop diabetes; three to seven times higher than normal people while obese individuals are 20 times higher (Joanne and Christopher 2018, Al-Goblan et al. 2014). The development of obesity augments the prevalence of type 2 diabetes. Numerous studies from various countries have ratified this association and reported a higher prevalence of diabetes among overweight and obese participants (Gregory et al. 2008).

It is relevant at this juncture to briefly refer to insulin resistance development induced by resistin. Resistin, an adipokine acts as a distinctive signaling molecule and has a significant role in inducing insulin resistance in obese patients. Type 2 diabetes is symbolized by target-tissue resistance to insulin. It is sturdily linked with obesity; albeit, the underlying mechanism by which increased adiposity induces insulin resistance remains still undetermined (Khalaf and Taegtmeyer 2012; Hampp et al. 2014).

It is found that an anti-resistin antibody can improve blood glucose with diet-induced obesity in mice. Furthermore, to treat the normal mice by recombinant resistin impedes glucose tolerance and insulin action. Insulin-triggered glucose uptake by adipocytes is increased by neutralization of resistin and is lessened by resistin treatment. Resistin is thus a hormone that possibly associates obesity with diabetes. Furthermore, glucose tolerance and insulin action are impaired with recombinant resistin to treat normal mice. Insulin-stimulated glucose uptake by adipocytes is reduced by resistin treatment and is enhanced by neutralization of resistin. Therefore, resistin signifies a hormone that possibly links obesity to diabetes (Claire et al. 2001).

Insulin resistance links to type 2 diabetes mellitus and obesity according to the plethora of studies. It was shown that quantitative control of the insulin-sensitive glucose transporters (Glut-4) and insulin receptors perse may subscribe to this disorder, nevertheless, these two factors are rather sufficient to clarify the degree of insulin resistance. This aspect has become apparent due to the development of just mild hyperinsulinemia in mice with a targeted mutation in the Glut-4 gene. Studies on post-receptor flaws in type 2 diabetes have currently focused on the intrinsic catalytic activity of the insulin receptor and downstream signaling events. A declined in tyrosine phosphorylation of both the insulin receptor (IR) and the insulin receptor substrate-1 (IRS-1) has been observed not only in animals but also among type 2 diabetes patients. Notably, this seems to take place in all of the major insulin-sensitive tissues. It is evident that the lessened signaling potential of the insulin receptor represents a significant element of this disease (Hotamisligil 1999).

8.3.3 Hypertension

Hypertension signifies systolic blood pressure more than 130 millimeters over diastolic 80 millimeters of mercury (mmHg), according to the American Heart Association (AHA) guidelines in 2017. Blood pressure is determined both by the quantity of blood heart pumps and the amount of arterial resistance to blood flow. Hypertension invariably leads to arterial narrowing. The association between blood pressure and body weight was reported in the 1960s and remained ambiguous till the mid-1980s it was clarified about the association between these two diseases by basic clinical and population-based research. Subsequent studies from different countries revealed a robust link between obesity and hypertension. Approximately 690 million people from all over the world have suffered from hypertension and nearly three million people die annually directly from the consequences of hypertension (Huang et al. 1998). It is known that excess weight puts pressure continuously on all organs of the body of the obese individual. As a result of this, obese individuals are beset with many health risks apart from hypertension (Gregory et al. 2008). Obesity epidemics have changed lifestyle and nutritional modification by increasing the intake of calories with reduced physical activity. It was reported that the increasing prevalence of hypertension is followed by increasing in the occurrence of obesity. Both hypertension and obesity are important public health problems globally. These diseases are highly connected, relevant and have increased rapidly since the year 2000 and it is projected that the number of hypertensive people will be increased to nearly 60% by the year 2025 (Kearney et al. 2005, Krzysztof 2006). It has been reported that among patients over 60 years of age, the association between systolic and diastolic blood pressure values and the body mass index is very significant (Praso et al. 2012).

Despite the use of anti-hypertensive drugs, the prevalence of hypertension remains increasing among obese patients (Gregory et al. 2008). As hypertension is linked with increasing body weight and any reduction in body weight will drop the blood pressure. Abdominal obesity was reported to be related to 2–3 times escalate the risk of hypertension. However, the risk of hypertension might be declined with a reduction of waist size to 102 cm among men and 88 cm among women (Okosun et al. 1999). Approximately 10% of weight-loss resulted in an average drop of 4.3/3.8 mmHg in blood pressure (Jayme 2018). Abdominal obesity and weight gain were regarded as one of the biomarkers of risk of hypertension. The incidence of hypertension-related directly to obesity was found in at least 75% of the American population (American Heart Association 2009). It was found when the weight is increased by more than 55 pounds among women, the comparative hazard of hypertension was three times more compared to those who increased by less than 10–22 pounds (John et al. 2015).

High blood pressure with resultant atherosclerosis may lead to develop heart attack, stroke or other health conditions such as aneurysm, weakened and narrowed blood renal and retinal vessels, heart failure, metabolic syndrome, impairment of memory or understanding and dementia. Obesity-related hypertension is often related to insulin resistance and glucose intolerance (Krzysztof 2006). Activation of the sympathetic nervous, and the renin-angiotensin systems with sodium retention describe the obesity-associated arterial hypertension among other abnormalities but this relation remains unexplored (Shu-Zhong et al. 2016).

8.3.4 Dementia

The brain has many distinct regions responsible for various functions. So, if there is any damage in a particular region, that region cannot carry out its functions normally. Dementia is caused by damaging the brain cells which leads to present a group of symptoms related to memory weakness or other thinking skills. Therefore, the ability to perform daily activities will be reduced such as to communicate with each other. Dementia has some symptoms including memory loss, communication and language, ability to focus and pay attention, judgment and visual perception. There are some non-modifiable risk factors for dementia such as age and genetics factors. However, researchers keep exploring the impact of other risk factors on brain health in order to prevent dementia.

Inflammation and oxidative stress may play the main roles in human health life. For example; inflammation is augmented systematically by both diabetes and obesity. Also, the brain has a significant influence on the inflammatory state of obesity/diabetes through numerous pathways. The pathologic changes of adiposopathy have a major influence on the brain structures and function which may be involved in dementing processes (Kiliaan et al. 2014, Gustafson 2008; Xu et al. 2011).

Several studies have reported that midlife obesity increases the risk of dementia in elderly people but the effect of obesity on dementia is controversial and the mechanisms of this association are still limited (Xu et al. 2011). People aged more than 65 years affected by dementia due to obesity were 6–10% but it remains poorly understood. Previous studies were conducted between 1995–2007 among older adults (40–80 years) showed that there was an association between the presence the risks of dementia and Alzheimer's disease (AD) with obesity. However, it was reported that Body mass index (BMI) and waist circumference (WC) was one of the predictors to develop dementia with a significant association (Beydoun et al. 2008).

Another study showed that after adjustment for potential confounders, midlife BMI was significantly related to increasing the risk of dementia, Alzheimer's disease (AD), vascular dementia as well as questionable dementia (Xu et al. 2011). Around 70–100% increase in the risk of developing dementia and Alzheimer's disease were attributed to obesity in mid-life. However, doing regular physical activities were reported to lower the risk of dementia, Alzheimer's disease and cognitive impairment compared to those exercising little (Gustafson 2006; Kivipelto et al. 2005; Whitmer et al. 2005; Rockwood and Middleton 2007).

8.3.5 Rheumatoid Arthritis

Rheumatoid arthritis (RA) is known as an autoimmune disease in which the body's immune system mistakenly attacks and damages the joints causing swelling and pain in and around the joints. This chronic inflammatory disorder also may damage the eyes, lungs, skin, heart and blood vessels. The joint damage has occurred on both sides of your body. If the left leg was affected and damaged, the right leg will probably be damaged as well.

Adults with Rheumatoid Arthritis (RA) who are obese often become disabled, experienced a more rapid decline in function and pain more quickly than people with RA who have a healthier weight. So, it is advisable for people living with RA to lose some pounds to do some exercises and using a proper diet. However, weight loss is also associated with worsening disability related to their frailty (Joshua et al. 2018). Conflict results were found reporting the association between obesity and RA. Some studies show that increasing the prevalence of obesity worldwide over the past several decades was related to develop the RA while other studies reported a negative association (George and Baker 2016).

It was consistently found that smoking was associated with escalating the risk of RA whereas obesity found in some previous studies could be associated with RA (Crowson et al. 2013; Myasoedova et al. 2010). Another meta-analysis study in 2015 demonstrated that there was a significant relationship between obesity and RA as obese individuals showed a significantly increased risk of RA after adjusting for the normal weight with a BMI less than 30 (RR = 1.25, 95% CI: 1.07 to 1.45) (Qin et al. 2015; Baodong et al. 2015; Deborah et al. 1997).

A prospective cohort study among women found that there was an increase in all RA risk among women who were overweight and obese and this association was stronger among those aged less than 55 years. Also, young women who overweight or obese reported increasing the seropositive and seronegative RA (Lu et al. 2014). Between 1985–2007, the incidence of RA increased by an elevated of 9.2/100,000 among women and 4.8/100,000 of this increment was accounted by obesity. This cohort study showed a significant association between the history of obesity with developing RA (Cynthia et al. 2013).

8.3.6 Dyslipidemia

Dyslipidemia is an abnormal amount of lipids in the blood that are too high or low. The elevation of lipids in the blood is the most common condition that can affect people worldwide due to their diet, lifestyle or prolonged elevation of insulin levels. Childhood obesity was reported as one of the risk factors of three cardiovascular including combined dyslipidemia (Freedman et al. 2007). Many studies found that increasing the incidences of both obesity and dyslipidemia leads to type 2 diabetes mellitus (T2DM) and cardiovascular disorders (CVDs) (Jayesh et al. 2015; Abhishek et al. 2015).

Dyslipidemia consists of increasing the level of low-density lipoprotein cholesterol (LDL-C), increasing the triglycerides (TG) and decreasing the high-density lipoprotein cholesterol (HDL-C). In dyslipidemia, the level of triglycerides (TG) is often between 100 and 400 mg/dL associated with obesity (Kwiterovich 2008). Besides, dyslipidemia was reported among obese (60–70%) and overweight patients (50–60%) as increasing in BMI, increasing the abnormalities in lipid levels (Bays 2013). The prevalence of increasing triglycerides and reduced HDL cholesterol levels was also reported among children and young adult obese and linked with the presence of other health conditions such as hypertension and diabetes (Skinner 2015; Gregory et al. 2008; Boudewijn et al. 2013).

The impact of obesity on lipid metabolism results in the location of the adipose tissue. For example, elevated visceral adipose tissue and trunk (especially upper trunk) subcutaneous adipose tissue are related to higher triglycerides and lower HDL cholesterol levels. On the other hand, elevated subcutaneous adipose tissue in the leg was related to lower triglycerides (Cameron et al. 2013; Grundy 2015).

8.3.7 Reproductive Disorders

Obesity is one of the major global issues associated with several reproductive disturbances. Obesity is related to several adverse reproductive consequences such as gestational diabetes mellitus, operative delivery, infertility, ovulation dysfunction, preeclampsia, preterm delivery, and fetal growth disorders (Elia et al. 2015). The mechanisms of obesity and its relation with fertility are complex and still not completely understood. It was reported that obesity may be associated with hyperandrogenism development, chronic anovulation, higher incidence of pre-eclampsia, subfecundity, infertility and oocyte abnormalities (Pasquali et al. 2003; Rich et al. 2002; Scialli 2006; Ramlau et al. 2007; Klenov and Jungheim 2014). Obese women are expected to practice pregnancy loss when pregnant and elevated miscarriage rates are found following natural conception, assisted conception and ovulation induction (Hamilton et al. 1992). In western society, obesity contributes to genetic and environmental factors to elevate the prevalence of reproductive disorders. More than 5000 infertile female patients in Australia have been reported with high BMI and this shows that BMI has been associated with reproductive performance (Clark et al. 1995). However, many obese women are fertile and have children. For example, not all obese women have polycystic ovary syndrome (PCOS), nor are all women with PCOS obese, but the two conditions are closely related. Therefore, the influence of obesity among female on human reproduction was primarily a subject of controversy (Bellver et al. 2006). Our discussion of obesity's relationship with reproductive disorders is brief in this chapter as it is covered more in detail in another chapter.

8.3.8 Respiratory Diseases

Respiratory diseases have become one of the major health problems and affect more than one-fifth of the population worldwide. Obesity is related to increasing the incidence of some health disorders and several abnormalities of respiratory function such as asthma, obesity hypoventilation syndrome, obstructive sleep apnea syndrome, chronic obstructive pulmonary disease, exertional dyspnea, aspiration pneumonia and pulmonary embolism (Magali et al. 2006). On the other hand, weight loss was related to reducing the pathogenicity of these complications and their treatment. However, the mechanisms of obesity's contribution to develop respiratory diseases remain unclear.

It was reported that these abnormalities are attributed to the accumulation of adipose tissue and the various cytokines are produced by adipocytes leading to lower lung volume due to a decrease in the chest wall. This will cause more restricted air entry which lowers the forced expiratory volume in 1 s (FEV1), forced vital capacity (FVC), functional residual capacity (FRC), expiratory reserve volume (ERV), residual volume (RV) and total lung capacity (TLC). Also, the function of the respiratory muscles deteriorates among obese people (Thiago et al. 2016; Christopher et al. 2010). Obesity was reported to have a clear association with dyspnea as it escalates the breathing due to reductions in both chest wall compliance and respiratory muscle strength. This may lead to an imbalance between the need for the respiratory muscles and their capacity to generate tension, which causes the perception of increased breathing effort (Sin et al. 2002; Weiner et al. 1998; Chlif et al. 2005).

Chronic obstructive pulmonary disease (COPD) is a worldwide health problem and consider as one of the leading causes of death particularly in the form of emphysema. It is associated with muscle wasting and weight loss which is occurred in advanced disease (Marquis et al. 2002). There is a reverse relationship between obesity and COPD as patients with COPD tend to have a more sedentary lifestyle, which would relate to develop obesity (Pitta et al. 2005). It is also hypothesized that obesity may be a protective factor in COPD (Celli et al. 2004). On the other hand, obesity was reported to be one of the risk factors for one of the sleeping disorders called Obstructive Sleep Apnea. It is a result of increasing the deposition of the fat tissue in the pharyngeal region and reducing operating lung volumes. This leads to lessen the upper airway caliber, adjusting airway configuration and elevate their collapsibility; airways are thus susceptible to repetitive closures during sleep (Series 2002). Moreover, it was revealed that around 70% of people with obstructive sleep apnea were from the obese group (Resta et al. 2001). Several studies showed that there is an association between obesity and many respiratory disorders. However, the mechanism of some disorders remains insufficient.

The respiratory problems are caused by obesity are preventable. The risk factors that might develop respiratory diseases can be prevented by avoiding stress factors, tobacco smoke, air pollution, pollutants, allergens, occupational agents, and other factors related to nutrition. So, we have to change our behavior, get an appropriate diet and do regular physical exercise. Our discussion of the impact of obesity on respiratory diseases is brief as it is covered more in detail in another chapter.

8.3.9 Renal Diseases

Renal diseases or kidney failure refers to the gradual loss of kidney ability and function. The kidneys filter and eliminate wastes and extra fluids from the blood, which are then secreted in the urine. When fluid, electrolytes and wastes are accumulated and elevated in the body; it will cause serious disorders in the advanced stage of kidney failure. Obesity is one of the risk factors to develop and accelerates the progression of renal disease (Praga et al. 2000; Bonnet et al. 2001; De Jong et al. 2002). Another study reported that obesity increases the risk of developing some major risk factors for chronic kidney disease (CKD) such as diabetes and hypertension, and it has a direct impact on the development of CKD and end-stage renal disease (ESRD) (Kramer et al. 2005; Esiki et al. 2004). When the intraglomerular pressure is increased, it can damage the kidney structure and increase the risk of developing CKD in the long term. However, the mechanisms to explain the relationship between obesity and CKD remain unclear and addressed.

In several large population-based studies showed that the glomerular hyperfiltration, elevated losing urinary albumin and progressive loss of renal function were caused by obesity (Kambham et al. 2019; Ribstein et al. 1995). Recent population-based studies have emphasized the relation between obesity and renal diseases. These studies showed a shred of evidence that obesity causes renal damage in apparently healthy subjects. It was reported that body mass index was related with urinary albumin excretion independently, presence, developing low estimated GFR (Foster et al. 2008), losing of estimated GFR rapidly over time (Lu et al. 2015) and incidence of ESRD (De Jong et al. 2002; Vivante et al. 2012). It was found that the association between BMI and impairment of renal function is obvious even among people without obesity (Bosma et al. 2004). Furthermore, elevated renal tubular sodium reabsorption impairs pressure natriuresis and plays a major role in introducing obesity hypertension (Obesity and Overweight Fact Sheet 2014). However, the fact that most obese/overweight persons never complaining of kidney disease as escalating their weight alone is not sufficient to cause kidney impairment (Bluher 2010).

8.3.10 Sleep-Disorders

Sleep apnea is a common disorder that may lead to more breathing pauses or shallow breaths during sleep. Recently, sleep disorders have been reported to have high prevalence particularly, among employed middle-aged people (Kronholm et al. 2008) and associated with body mass index (BMI). Sleep apnea may cause accumulating more fat around the neck which leads to narrow the airway, causing breathing difficulty. Many findings from different researches show the relationship between sleep disorders and gaining weight. In contrast, gaining weight is presented due to insufficient sleep. Therefore, the relationship can be in two ways of direction.

Several studies demonstrated the relationship between sleep period with BMI and gaining weight (Stranges et al. 2008; López-García et al. 2008; Lauderdale et al.

2009). Most of the available studies in the literature showed that there was an association between obesity and sleep disorders (Fogelholm et al. 2007; Van den Berg et al. 2008; Lauderdale et al. 2009). Furthermore, a study was conducted among men and women in the middle-aged showed that suffering from falling asleep, waking up many times per night and suffering from staying asleep were related with gaining weight among women but not among men (Lyytikäinen et al. 2011). However, some other studies reported that there were no relationships between sleep disorders and BMI (Owens and Matthews 1998; Björkelund et al. 2005). Therefore, these results from the literature are inconsistent might be attributed to the differences in sample, sampling or sample size.

8.3.11 Cancers

Cancer also called malignancy which is a collection of related diseases with abnormal cell growth which can be spread into surrounding tissues. Cancer has types of cancers such as lung cancer, skin cancer, breast cancer, colon cancer, prostate cancer, and lymphoma with different symptoms according to its type. Consistent findings from different researches (mostly from large cohort studies and observational studies) involving millions of participants revealed that there was a strong relationship between obesity and cancer. However, findings from observational studies could not identify the causality of obesity to develop cancers due to the differences between obese and underweight people concerning other than their body fat.

There are various explanations about the mechanisms on how obesity can elevate the risks of some cancers. It was explained that fat cells are producing some hormones that lead to stimulate or inhibit cell growth such as adipokines. Adipose tissue has a link to develop cancers as it produces additional amounts of estrogen which have an association with escalated risks of breast, ovarian, endometrial and some other cancers. Furthermore, chronic inflammation may lead to DNA damage which causes cancer among obese. Hyperinsulinemia or insulin resistance also reported promoting to development of colon, kidney, prostate, and endometrial cancers (Gallagher and LeRoith 2015; Seo et al. 2015).

Plentiful studies have reported the relation between various cancers and obesity. A meta-analysis study revealed that BMI has a positive association with malignant melanoma among men only, leukemia, multiple myeloma and thyroid cancer (Renehan et al. 2008). In UK, it was revealed that around 17,294 cancer cases reported in 2010 were attributed to overweight and obesity (Parkin and Boyd 2011). Other studies showed more evidences for obesity to have a link with cancer such as prostatic cancer has a statistical significant with BMI (Bergström et al. 2001), liver cancer and gastric cardia have an association with obesity (Larsson and Wolk 2007; Calle and Kaaks 2004).

It is evident that obesity among cancer survivors reported from different studies for certain cancers such as prostate, colorectal or breast cancer. It was reported that obesity can affect and deteriorate various aspects of cancer survivorship, including cancer progression, quality of life, cancer recurrence and prognosis (Calle et al. 2003; Schmitz et al. 2013). More explanation about the relationship between cancer and obesity will be discussed in detail in a book entitled Obesity and Cancer Current Concepts by Kannan Kutty and Muhamed Othman in 2013.

8.4 Summary

In this chapter, it was discussed and addressed some comorbidities that related with obesity and highlighted its impact on many other diseases or disorders. It was shown that obesity has been linked with some serious and common diseases such as diabetes, hypertension and some cardiovascular diseases which make it one of the major health problems nowadays. There are many other diseases that have associated with obesity and difficult to discuss here in this chapter. However, some of these diseases are discussed in other chapters in this book by other authors. This chapter is written in the hope that it will prove the association between obesity and some other diseases and would be highly beneficial to the academicians, researchers and students.

References

Abhishek G, Vinod KT, Sunil Kumar V (2015) Evaluation of pattern of dyslipidemia and its association with obesity in type 2 diabetes mellitus. Int J Contemp Med Res 2(5): 1432-1435

Al-Goblan A, Al-Alfi M, Khan M (2014) Mechanism linking diabetes mellitus and obesity. Diabetes Metab Syndr Obes. 4;7:587-91. doi: 10.2147/DMSO.S67400. PMID: 25506234; PMCID: PMC4259868

Alpert MA, Terry BE, Mulekar M (1997) Cardiac morphology and left ventricular function in normotensive morbidly obese patients with and without congestive heart failure, and effect of weight loss. Am J Cardiol 80: 736-740

Alpert MA (2001) Obesity cardiomyopathy: pathophysiology and evolution of the clinical syndrome. Am J Med Sci 321: 225-236

American Heart Association, Overweight and Obesity Statistics – 2009 Update. 2009

Ashrafian H, le Roux CW, Darzi A, Athanasiou T (2008) Effects of bariatric surgery on cardiovascular function. Circulation 118(20): 2091-2102

Baodong Q, Min Y, Haitao F, Ning M, Tingting W, Qingqin T, Zhide H, Yan L, Zaixing Y Renqian Z (2015) Body mass index and the risk of rheumatoid arthritis: a systematic review and dose-response meta-analysis. Arthritis Res Ther 17: 86

Bays HE (2013) Obesity, adiposity, and dyslipidemia: a consensus statement from the National Lipid Association. J Clin Lipidol 7(4): 304-383.

Bellver J, Busso C, Pellicer A, Remohi J, Simon C (2006) Obesity and assisted reproductive technology outcomes. Reprod Biomed Online 12: 562-568

Bergström A, Pisani P, Tenet V, Wolk A, Adami HO (2001) Overweight as an avoidable cause of cancer in Europe. Int J Cancer. 1;91(3):421-30. doi: 10.1002/1097-0215(200002)9999:9999<::aid-ijc1053>3.0.co;2-t

Björkelund C, Bondyr-Carlsson D, Lapidus L, Lissner L, Månsson J, Skoog I (2005) Sleep disturbances in midlife unrelated to 32-year diabetes incidence. Diabetes Care 28: 2739-2744

Bluher M (2010) The distinction of metabolically "healthy" from "unhealthy" obese individuals. Curr Opin Lipidol 21: 38-43

Bonnet F, Deprele C, Sassolas A (2001) Excessive body weight as a new independent risk factor for clinical and pathological progression in primary IgA nephritis. Am J Kidney Dis 37: 720-727

Bosma RJ, Homan van der Heide JJ, Oosterop EJ, De Jong PE, Navis GJ (2004) Body mass index is associated with altered renal hemodynamics in non-obese healthy subjects. Kidney Int 65: 259-265

Boudewijn K, Jan Willem FE, Manuel CC (2013) Dyslipidemia in Obesity: Mechanisms and Potential Targets. Nutrients 5, 1218-1240

Branca F (2010) Obesity and Public Health. Global History Seminars Geneva. Department of Nutrition for Health and Development, WHO

Calle E, Kaaks R (2004) Overweight, obesity and cancer. Epidemiological evidence and proposed mechanisms. Nat Rev Cancer 4: 579-591

Calle EE, Rodriguez C, Walker-Thurmond K, Thun MJ (2003) Overweight, obesity, and mortality from cancer in a prospectively studied cohort of U.S. adults. N Engl J Med 348(17): 1625-1638

Cameron AJ, Magliano DJ, Soderberg S (2013) A systematic review of the impact of including both waist and hip circumference in risk models for cardiovascular diseases, diabetes and mortality. Obes Rev 14(1): 86-94

Carl JL, Richard VM, Hector OV (2009) Obesity and Cardiovascular Disease. Risk Factor, Paradox, and Impact of Weight Loss. J Am Coll Cardiol (53)21: 1925-1932

Celli BR, Cote CG, Marin JM, et al (2004) The body-mass index, airflow obstruction, dyspnea, and exercise capacity index in chronic obstructive pulmonary disease. N Engl J Med 350: 1005-1012

Center for Disease Control and Prevention (CDC) (2019) Living with Diabetes: http://www.cdc.gov/diabetes/living/index.html

Chlif M, Keochkerian D, Mourlhon C (2005) Noninvasive assessment of the tension-time index of inspiratory muscles at rest in obese male subjects. Int J Obes (Lond) 29: 1478-1483

Christopher Z, Helen L, Ian M, Himender M (2010) Obesity and respiratory diseases. Int J Gen Med 3: 335-343

Claire MS, Shannon TB, Savitha B, Elizabeth JB, Ronadip RB, Christopher MW, Hiralben RP, Rexford SA, Mitchell A (2001) The hormone resistin links obesity to diabetes. Nature 409

Clark AM, Ledger W, Galletly C (1995) Weight loss results in significant improvement in pregnancy and ovulation rates in anovulatory obese women. Hum Reprod 10: 2705-2712

Crowson CS, Matteson EL, Davis JM, Gabriel SE (2013) Contribution of obesity to the rise in incidence of rheumatoid arthritis. Arthritis Care Res (Hoboken) 65(1): 71-77

Cynthia SC, Eric LM, John MD, Sherine EG (2013) Contribution of obesity to the rise in incidence of rheumatoid arthritis. Arthritis Care Res (Hoboken) 65(1): 71-77

De Jong PE, Verhave JC, Pinto-Sietsma SJ, Hillege HL (2002) Obesity and target organ damage: the kidney. Int J Obes Relat Metab Disord 26(S4): S21-S24

Deborah PM, Symmons CR, Bankhead, Beverley JH, Paul B, Alan JS, Elizabeth MB, David GI, Scott (1997) Blood transfusion, smoking, and obesity as risk factors for the development of rheumatoid arthritis. Results from a primary care-based incident case-control study in Norfolk, England. Arthritis Sr Rheumatism 40 (11): 1955-1961

Elia EM, Bazzano MV, Quintana R, Paz DA (2015) Reproductive disorders in obesity. Integrative Obesity and Diabetes 1(1): 20-25

Esiki K, Ikemiya Y, Kinjo K, Inoue T, Esiki C, Takishita S (2004) Body mass index and the risk of development of end-stage renal disease in a screened cohort. Kidney Int 65: 1870-1876

Fogelholm M, Kronholm E, Kukkonen-Harjula K, Partonen T, Partinen M, Härmä M (2007) Sleep-related disturbances and physical inactivity are independently associated with obesity in adults. Int J Obes 31: 1713-1721

Foster MC, Hwang SJ, Larson MG (2008) Overweight, obesity, and the development of stage 3 CKD: the Framingham heart study. Am J Kidney Dis 52: 39-48

Freedman DS, Khan LK, Serdula MK, Galuska DA, Dietz WH (2002) Trends and correlates of class 3 obesity in the United States from 1990 through 2000. JAMA 288(14): 1758-1761

Freedman DS, Mei Z, Srinivasan SR, Berenson GS, Dietz WH (2007) Cardiovascular risk factors and excess adiposity among overweight children and adolescents: the Bogalusa heart study. J Pediatr 150(1):12-17

Galbete C, Toled J, Martinez MA, Martinez JA, Guillen GF (2013) Lifestyle factors modify obesity risk linked to PPARG2 and FTO variants in an elderly population: a cross-sectional analysis in the SUN Project. Genes Nutr 8: 61-67

Gallagher EJ, LeRoith D (2015) Obesity and diabetes: The increased risk of cancer and cancer-related mortality. Physiol Rev 95(3): 727-748

Gregg EW, Cheng YJ, Cadwell BL, et al (2005) Secular trends in cardiovascular disease risk factors according to body mass index in US adults. JAMA 293(15): 1868-1874

Gregory LB, Alain GB, Steven S, Russell T, Karol EW, Roger SB, Hyoju C, Mercedes RC (2008) The impact of obesity on cardiovascular disease risk factors and subclinical vascular disease: The multi-ethnic study of atherosclerosis. Arch Intern Med 168(9): 928-935

Grundy SM (2015) Adipose tissue and metabolic syndrome: too much, too little or neither.Eur J Clin Invest 45(11): 1209-1217

Gustafson D (2008) A life course of adiposity and dementia. Eur J Pharmacol 585: 163-175

Gustafson D (2006) Adiposity indices and dementia. Lancet Neurol 5: 713-720

Hamilton D, Kiddy D, Watson H, Paterson C, Franks S (1992) Association of moderateobesity with a poor pregnancy outcome in women with polycystic ovary syndrome treated with low dose gonadotrophin. Br J Obstet Gynaecol 99: 128-131

Hampp C, Borders-Hemphill V, Moeny D, Wysowski D (2014) Use of antidiabetic drugs in the U. S., 2003-2012. Diabetes Care. 37(5):1367-74. doi: 10.2337/dc13-2289. Epub 2014 Mar 12. PMID: 24623020

Haslam DW, James WP (2005) Obesity. Lancet 366: 1197-1209

Hotamisligil GS (1999) The role of TNF alpha and TNF receptors in obesity and insulin resistance. J Intern Med 245(6): 621-625

Huang Z, Willett WC, Manson JE (1998) Body weight, weight change, and risk for hypertension in women. Ann Intern Med 128: 81-88

Jayesh S, Ankna S, Frenny S, Sunil T, Nutan N, Navneet S, Premal T, Rama V. The association of dyslipidemia and obesity with glycated hemoglobin. Clin Diabetes Endocrinol 20151: 6

Jaymee D (2018) Hypertension and obesity: How Weight-loss affects hypertension. The obesity action coalition. https://www.obesityaction.org/community/article-library/hypertension-and-obesity-how-weight-loss-affects-hypertension/#; assessed on 8th December 2018

Joanne ZR, Christopher DS (2018) Obesity and Type 2 Diabetes. https://www.obesityaction.org/community/article-library/obesity-and-type-2-diabetes/; assessed on 20th July 2019

John EH, Jussara MC, Alexandre AS, Zhen W, Michael EH (2015) Obesity-induced hypertension. Circulation Res 116(6): 991-1006

Joshua FB, Bryant RE, Ted RM, Harlan S, Grant WC, Brian CS, Michael DG, Liron C, Kaleb M (2018) Obesity, Weight Loss, and Progression of Disability in Rheumatoid Arthritis. Arthritis care & Res 70 (12): 1740-1747

June MC, Eric BR, Graham AC, Meir JS, Walter CW (1994) Obesity, fat distribution, and weight gain as risk factors for clinical diabetes in men. Diabetes Care 17(9): 961-969

Kambham N, Markowitz GS, Valeri AM, Lin J, D'Agati VD (2019) Obesity-related glomerulopathy: an emerging epidemic. Kidney Int 59: 1498-1509

Kasuga M (2006) Insulin resistance and pancreatic beta cell failure. J Clin Invest. 116(7):1756-60. doi: 10.1172/JCI29189. PMID: 16823472; PMCID: PMC1483164

Kearney PM, Whelton M, Reynolds K, Muntner P, Whelton PK, He J (2005) Global burden of hypertension: analysis of worldwide data. Lancet 365: 217-223

Kenchaiah S, Evans JC, Levy D (2002) Obesity and the risk of heart failure. N Engl J Med 347: 305-313

Khalaf KI, Taegtmeyer H (2012) After avandia: the use of antidiabetic drugs in patients with heart failure. Tex Heart Inst J 39(2): 174-178

Kiliaan AJ, Arnoldussen IAC, Gustafson DR (2014) Adipokines: A link between obesity and dementia? Lancet Neurol 13: 913-923.

Kivipelto M et al (2005) Obesity and vascular risk factors at midlife and the risk of dementia and Alzheimer disease. Arch Neurol. 62:1556-1560.

Klenov VE, Jungheim ES (2014) Obesity and reproductive function: a review of the evidence. Curr Opin Obstet Gynecol 26: 455-460

Kramer H, Luke A, Bidani A, Cao G, Cooper R, McGee D (2005) Obesity and prevalent and incident CKD: the hypertension detection and follow-up program. Am J Kidney Dis 46: 587-594

Kronholm E, Partonen T, Laatikainen T, Peltonen M, Härmä M, Hublin C (2008) Trends in self-reported sleep duration and insomnia-related symptoms in Finland from 1972 to 2005: a comparative review and re-analysis of Finnish population samples. J Sleep Res 17: 54-62

Krzysztof N (2006) Obesity and hypertension-the issue is more complex than we thought. Nephrol Dial Transplant 21(2): 264-267

Kuller L, Borhani N, Furberg C, Gardin J, Manolio T, O'Leary D, Psaty B, Robbins J (1994) Prevalence of subclinical atherosclerosis and cardiovascular disease and association with risk factors in the Cardiovascular health study. Am J Epidemiol 139: 1164-1179

Kwiterovich PO Jr (2008) Recognition and management of dyslipidemia in children and adolescents. J Clin Endocrinol Metab 93(11): 4200-4209

Kwok CY, Wong CNY, Yau MYC, Yu PHF, Au ALS, Poon CCW, Seto SW, Lam TY, Kwan YW, Chan SW (2010) Consumption of dried fruit of Crataeguspinnatifida (hawthorn) suppresses high-cholesterol diet-induced hypercholesterolemia in rats. J Funct Foods 179-186

Larsson B, Svardsudd K, Welin L, Wilhelmsen L, Bjorntorp P, Tibblin G (1984) Abdominal adipose tissue distribution, obesity, and risk of cardiovascular disease and death: 13 year follow up of participants in the study of men born in 1913. Br Med J 288: 1401-1407

Lars F, Lone JH, Peter V (2005) Overweight and obesity as risk factors for atrial fibrillation or flutter: The Danish Diet, Cancer, and Health Study. Am J Med 118(5): 489-495

Larsson SC, Wolk A (2007) Overweight, obesity and risk of liver cancer: a meta-analysis of cohort studies. Br J Cancer 97: 1005-1008

Lauderdale DS, Knutson KL, Rathouz PJ, Yan LL, Hulley SB, Liu K (2009) Cross-sectional and longitudinal associations between objectively measured sleep duration and body mass index. Am J Epidemiol 170: 805-813

Lauer MS, Anderson KM, Kannel WB, Levy D (1991) The impact of obesity on left ventricular mass and geometry: The Framingham heart study. JAMA 266(2): 231-236

Ledwozyw A, Michalak J, Stepień A, Kadziołka A (1986) The relationship between plasma triglycerides, cholesterol, total lipids and lipid peroxidation products during human atherosclerosis. Clin Chim Acta 155: 275-283

Lewis L, Louis JA, Lawrence JB, Valerie B, Leon II, Donald L, James S (2013) Obesity-related hypertension: pathogenesis, cardiovascular risk, and treatment. J Clin Hypertens 15(1)

Life JS (2014) The Life Plan Diet: How Losing Belly Fat is the Key to Gaining a Stronger, Sexier, Healthier Body. Simon and Schuster; New York, NY.

López-García E, Faubel R, León-Muñoz L, Zuluaga MC, Banegas JR, Rodríguez-Artalejo F (2008) Sleep duration, general and abdominal obesity, an weight change among the older adult population of Spain. Am J Clin Nutr 87: 310-316

Lu B, Hiraki LT, Sparks JA (2014) Being overweight or obese and risk of developing rheumatoid arthritis among women: a prospective cohort study. Ann Rheum Dis 73: 1914-1922

Lu JL, Molnar MZ, Naseer A, Mikkelsen MK, Kalantar-Zadeh K, Kovesdy CP (2015) Association of age and BMI with kidney function and mortality: a cohort study. Lancet Diabetes Endocrinol 3:704-714

Luma A, Anwar A (2011) Relationships between obesity and cardiovascular diseases in four southern states and Colorado. J Health Care Poor Underserved 22(S4): 61-72

Lyytikäinen P, Lallukka T, Lahelma E, Rahkonen O (2011) Sleep problems and major weight gain: a follow-up study. Int J Obes 35: 109-114

Beydoun MA, Beydoun HA, Wang Y (2008) Obesity and central obesity as risk factors for incident dementia and its subtypes: a systematic review and meta-analysis. Obes Rev 9(3): 204-218

Magali P, Mariève D, Geneviève CM, Vicky D, Frédéric S, Louis-Philippe B, Angelo T, François M (2006) The effect of obesity on chronic respiratory diseases: pathophysiology and therapeutic strategies. CMAJ 174(9): 1293

Marquis K, Debigaré R, LeBlanc P (2002) Mid-thigh muscle cross-sectional area is a better predictor of mortality than body mass index in patients with COPD. Am J Respir Crit Care Med 166: 809-813

Matsuzawa Y (2005) Metabolic syndrome - definition and diagnostic criteria in Japan. J Atheroscler Thromb 12: 301

Messerli FH, Nunez BD, Ventura HO, Snyder DW (1987) Overweight and sudden death: increased ventricular ectopy in cardiomyopathy of obesity. Arch Intern Med 147: 1725-1728

Messerli FH, Ventura HO, Reisin E (1982) Borderline hypertension and obesity: two prehypertensive states with elevated cardiac output. Circulation 66: 55- 60

George M and Baker J (2016) The Obesity Epidemic and Consequences for Rheumatoid Arthritis Care. Curr Rheumatol Rep 18(1): 6

Myasoedova E, Crowson CS, Kremers HM, Therneau TM, Gabriel SE (2010) Is the incidence of rheumatoid arthritis rising?: results from Olmsted County, Minnesota, 1955-2007. Arthritis Rheum 62(6): 1576-1582

Nestle M, Jacobson MF (2000) Halting the obesity epidemic: A public health policy approach. Public Health Rep 115: 12-24. DOI: https://doi.org/10.1093/phr/115.1.12

Obesity and Overweight Fact Sheet No. 311 (2014) http://www.who.int/mediacentre/factsheets/fs311/en/.2014.

Okosun IS, Prewitt TE, Cooper RS (1999) Prevalence and attributable risk of hypertension. J Hum Hypertens 13: 425430

Oren A, Vos LE, Uiterwaal CS, Gorissen WH, Grobbee DE, Bots ML (2003) Change in body mass index from adolescence to young adulthood and increased carotid intima-media thickness at 28 years of age: The atherosclerosis risk in young adults study. Int J Obes Relat Metab Disord 27 (11): 1383-1390

Ostlund, RE Jr, Staten, M, Kohrt, WM (1990) The ratio of waist-to-hip circumference, plasma insulin level, and glucose intolerance as independent predictors of the HDL2 cholesterol level in older adults. N Engl J Med 322: 229

Owens JF, Matthews KA (1998) Sleep disturbance in healthy middle-aged women. Maturitas 30: 41-50

Parkin DM, Boyd L (2011) Cancers attributable to overweight and obesity in the UK in 2010. Br J Can. 105: S34 - S37

Pasquali R, Pelusi C, Genghini S, Cacciari M, Gambineri A (2003) Obesity and reproductive disorders in women. Hum Reprod Update 9(4): 359-372

Pitta F, Troosters T, Spruit MA (2005) Characteristics of physical activities in dailylife in chronic obstructive pulmonary disease. Am J Respir Crit Care Med 171: 972-977

Poirier P, Giles TD, Bray GA (2006) Obesity and cardiovascular disease: pathophysiology, evaluation, and effect of weight loss: an update of the 1997 American Heart Association scientific statement on obesity and heart disease from the obesity committee of the council on nutrition, physical activity, and metabolism. Circulation 113: 898-918

Praga M, Hernandez E, Herrero JC (2000) Influence of obesity on the appearance of proteinuria and renal insufficiency after unilateral nephrectomy. Kidney Int 58: 2111-2118

Praso S, Jusupovic F, Ramic E, Gledo I, Ferkovic V, Novakovic B, Hadzovic E (2012). Obesity as a risk factor for artherial hypertension. Materia socio-medica, 24(2), 87–90. https://doi.org/10.5455/msm.2012.24.87-90

Qin B, Yang M, Fu H, Ma N, Wei T, Tang Q, Hu Z, Liang Y, Yang Z, Zhong R (2015) Review Body mass index and the risk of rheumatoid arthritis: a systematic review and dose-response meta-analysis. Arthritis Res Ther 29(17): 86

Racette SB, Deusinger SS, Strube MJ, Highstein GR, Deusinger RH (2005) Weight changes, exercise, and dietary patterns during freshman and sophomore years of college. J Am Coll Health 53: 245-251
Ramlau CH, Thulstrup AM, Nohr EA, Bond JP, Sarensen TIA (2007) Subfecundityin overweight and obese couples. Hum Reprod 22: 1634-1637
Renehan AG, Tyson M, Egger M, Heller RF, Zwahlen M (2008) Body-mass index and incidence of cancer: a systematic review and meta-analysis of prospective observational studies. Lancet 371: 569-578
Resta O, Foschino-Barbaro MP, Legari G (2001) Sleep-related breathing disorders, loud snoring and excessive daytime sleepiness in obese subjects. Int J Obes Relat Metab Disord 25: 669-675
Ribstein J, du Cailar G, Mimran A (1995) Combined renal effects of overweight and hypertension. Hypertension 26: 610-615
Rich JW, Spiegelman D, Garland M, Hertzmark E, Hunter DJ (2002) Physical activity, body mass index, and ovulatory disorder infertility. Epidemiol 13: 184-190
Rockwood K, Middleton L (2007) Physical activity and the maintenance of cognitive function. Alzheimer Dement 3: S38-S44
Rosenbaum M, Leibel RL (1998) The physiology of body weight regulation: Relevance to the etiology of obesity in children. Pediatrics 101: 525-539
Sadiya SK, Hongyan N, John TW, Norrina A, Mercedes C, Jarett DB, Ranya NS, Donald ML (2018) Association of body mass index with lifetime risk of cardiovascular disease and compression of morbidity. JAMA Cardiol 3(4): 280-287
Schmitz KH, Neuhouser ML, Agurs-Collins T (2013) Impact of obesity on cancer survivorship and the potential relevance of race and ethnicity. J Natl Cancer Inst 105(18): 1344-1354
Scialli AR (2006) Teratology public affairs committee position paper: Maternal obesity and pregnancy. Birth Defects Res A Clin Mol Teratol 76: 73-77
Seo BR, Bhardwaj P, Choi S (2015) Obesity-dependent changes in interstitial ECM mechanics promote breast tumorigenesis. Sci Transl Med 7(301): 301
Sériès F (2002) Upper airway muscles awake and asleep. Sleep Med Rev 6: 229-242
Shu-Zhong J, Wen L, Xue-Feng Z, Hong-Yun R, Yi L (2016) Obesity and hypertension (Review). Exp Ther Med 12: 2395-2399
Sin DD, Jones RL, Man SF (2002) Obesity is a risk factor for dyspnea but not for air flow obstruction. Arch Intern Med 162: 1477-1481
Skinner AC (2015) Cardiometabolic risks and severity of obesity in children and young adults. N Engl J Med 373(14): 1307-1317
Skyler JS, Oddo C (2002) Diabetes trends in the USA. Diabetes Metab Res Rev 18(S3): S21-S26
Stranges S, Cappuccio FP, Kandala NB, Miller MA, Taggart FM, Kumari M, Ferrie JE, Shipley MJ, Brunner EJ, Marmot MG (2008) Cross-sectional versus prospective associations of sleep duration with changes in relative weight and body fat distribution: the Whitehall II Study. Am J Epidemiol. 1;167(3):321-9. doi: 10.1093/aje/kwm302. Epub 2007 Nov 15. PMID: 18006903; PMCID: PMC3206317
Subhija P, Fatima J, Enisa R, Ibrahim G, Vesna F, Budimka N, Emir H (2012) Obesity as a risk factor for artherial hypertension. Mat Soc Med 24(2): 87-90
Thiago TM, Rogério R, Cláudia HC, Agnaldo JL (2016) Obesity: Systemic and pulmonary complications, biochemical abnormalities, and impairment of lung function. Multidiscip Respir Med 11: 28
Thomas F, Bean K, Pannier B, Oppert JM, Guize L, Benetos A (2005) Cardiovascular mortality in overweight subjects: the key role of associated risk factors. Hypertension 46: 654-659
Van den Berg JF, Neven AK, Tulen JHM, Hofman A, Witteman JCM, Miedema HME (2008) Actigraphic sleep duration and fragmentation are related to obesity in the elderly: the Rotterdam Study. Int J Obes 32: 1083-1090
Vivante A, Golan E, Tzur D (2012) Body mass index in 1.2 million adolescents and risk for end-stage renal disease. Arch Intern Med 172:1644-1650

Wanahita N, Messerli FH, Bangalore S (2008) Atrial fibrillation and obesity-results of a meta-analysis. Am Heart J 155: 310-315

Wang TJ, Parise H, Levy D (2004) Obesity and the risk of new-onset atrial fibrillation. JAMA 292: 2471-2477

Wei M, Gaskill SP, Haffner SM, Stern MP (1997) Waist circumferences the best predictor of non-insulin dependent diabetes mellitus (NIDDM) compared to body mass index, waist/hip ratio over other anthropometric measurements in Mexican Americans: A 7-year Prospective Study. Obesity Res 5: 16-23

Weiner P, Waizman J, Weiner M (1998) Influence of excessive weight after gastroplasty for morbid obesity on respiratory muscle performance. Thorax 53: 39-42

Whitmer RA, Gunderson EP, Barrett-Connor E, Quesenberry CP Jr, Yaffe K (2005) Obesity in middle age and future risk of dementia: a 27 year longitudinal population based study. BMJ. 11;330(7504):1360. doi: 10.1136/bmj.38446.466238.E0

Xu WL, Atti AR, Gatz M, Pedersen NL, Johansson B, Fratiglioni L (2011) Midlife overweight and obesity increase late-life dementia risk. Neurology 76(3): 1568-1574

Imaging as a Tool for Measuring Body Fat

9

Marymol Koshy, Mazuin Mohd Razalli, Asita Elengoe, and Methil Kannan Kutty

Abstract

Obesity is defined as a condition of abnormal or excessive fat accumulation that may affect health. It is a pan-global epidemic disease and has been escalating during the past two decades. Obesity is the major risk for a variety of chronic diseases such as diabetes, cardiovascular diseases and cancer. People think that obesity causes a problem in high-income countries such as the United States of America, the United Kingdom, etc. However, the obesity problem now arises dramatically in low-income countries too such as South Africa, India, etc. Body weight is the result of genetic defect, metabolism, behaviour, culture and environment. There are many methods for assessing body composition. Abdominal fat is distributed into two main compartments which have different metabolic characteristics: subcutaneous and visceral adipose tissue. Measuring visceral adipose tissue is extremely important as it is closely associated with metabolic syndrome eventually contributing to morbidity and mortality. In this chapter, we will discuss imaging, computed tomography, magnetic resonance imaging and ultrasound.

M. Koshy
Department of Radiology, Faculty of Medicine, UiTM, Shah Alam, Malaysia

M. M. Razalli
Department of Radiology, Faculty of Medicine, Universiti Teknologi MARA, UiTM, Shah Alam, Malaysia

A. Elengoe
Biotechnology Department, Faculty of Science, Lincoln University College, Petaling Jaya, Selangor, Malaysia

M. K. Kutty (✉)
Department of Pathology, Faculty of Medicine, Lincoln University College, Petaling Jaya, Selangor, Malaysia

M. K. Kutty, A. Elengoe (eds.), *Obesity and its Impact on Health*,
https://doi.org/10.1007/978-981-33-6408-0_9

Keywords

Obesity · Imaging · Body fat

9.1 Introduction

There are many methods for assessing body composition. Some of these are referred to as laboratory methods as they are only available in clinical and research settings. The common laboratory methods include bioelectrical impedance (BIA), dual-energy X-ray absorptiometry (DXA), densitometry obtained from underwater weighing or air displacement plethysmography, and hydrometry from isotope dilution. Sophisticated methods such as computed tomography scan (CT scan), dual-energy X-ray absorptiometry (DXA) and magnetic resonance imaging (MRI) are also called 'reference measurements' as they are only used in research to validate body measurements. All of these methods have pros and cons.

9.2 Imaging Central Obesity

Abdominal obesity also is known as central obesity. It is an important component of the cardiometabolic syndrome and has a strong association with stroke (Chen et al. 2006), coronary heart disease (Cabrera et al. 2007) and overall mortality (Gami et al. 2007; Jacobs et al. 2010).

Clinical measurements of the waist are simple and convenient however it cannot differentiate fat into different compartments. The abdominal fat compartment is divided into subcutaneous and visceral fat. To investigate the pathophysiology of obesity accurate measurements of visceral and subcutaneous fat are essential and most of the current research focuses on partitioning adipose tissue. Cardiometabolic risks are found to be associated more with visceral adipose tissue rather than subcutaneous adipose tissue. The ratio between visceral and subcutaneous fat is essential in predicting cardiometabolic risk. Direct imaging methods such as CT, MRI and ultrasound are available. Currently, CT and MRI are considered to be the most effective techniques for determining whole-body fat mass, lean muscle mass and bone mass. However, these methods are only used in research settings due to ethical and practical issues.

Numerous factors contribute to the development of obesity. However, overconsumption of food and physical inactivity are the most common causes of obesity. Bodyweight is the result of genetic defects, metabolism, behaviour, culture and environment.

Overeating induces an increase of weight; especially for high consumption of fat food. Food consists of high fat or sugar such as fast food (KFC, McDonald, Burger King and Pizza Hut), fried food (French fries, hotdog, and nuggets) and sweets have energy density. Even a small quantity of food has high calories. Studies have established that diets high in fat lead to weight gain. According to some researchers, simple carbohydrates such as sugars, fructose, desserts, soft drinks, beer, chocolates, potato chips lead to weight gain because they are absorbed into the blood circulation very quickly compared to complex carbohydrates like pasta, spaghetti, rice, vegetables, fruits. Hence, higher levels of hormone insulin are required for the metabolism of simple carbohydrates than complex carbohydrates. The elevated hormone insulin release induces weight gain.

Besides, studies demonstrated that people who consume a small amount of food 4 or 5 times daily have lower cholesterol levels and stable blood sugar levels compared to those who eat a large amount of food 2 or 3 times daily. This is because the pancreas releases a low level of hormone insulin for small frequent meals, while the high amount of insulin is required to break down the large meal.

Furthermore, people who are active burn more calories than inactive people. It has been proven that physical inactivity is strongly related to weight gain in both men and women. Regular exercise reduces the fats in the body.

Medications such as antidepressants, anticonvulsants (carbamazepine and valproate), diabetes (insulin, sulfonylureas and thiazolidinediones), certain hormones (oral conceptive and corticosteroids), high blood pressure medications and antihistamines can be responsible for weight gain. It is of interest that psychological factors also contribute to obesity. Emotions such as anger, boredom, sadness, stress or anger influence eating habits. In addition, hypothyroidism, insulin resistance, polycystic ovary syndrome (PCOS), and Cushing's syndrome diseases also cause obesity.

Moreover, additional factors such as inadequate monitory circumstances to buy healthy food or lack of safe places for jogging and walking can increase the risk of obesity. Imbalance of hormones also leads to weight gain, especially during pregnancy and menopause.

9.3 Computed Tomography

Computed tomography (CT) has been found to be an optimal method for the precise assessment of intra-abdominal fat. CT can provide a ratio of intra-abdominal fat to extra -abdominal fat. CT produces cross-sectional images of the body by sending a beam of photons from an x-ray tube towards a detector. The beam rotates around the subject and data is collected, stored and applied to complex algorithms to build images that evaluate body composition (Fig. 9.1). As only a few axial images are required, the overall radiation dose from this method is minimal. CT is non-invasive however it is limited by radiation exposure and the cost of the equipment is expensive. Two factors that have to be taken into consideration are the maximum weight the CT table allows and the girth of the patient. Always subtract 15–18 cm of table width from gantry aperture, which means a 90 cm gantry aperture will only allow a patient with 72–78 cm girth.

Using a CT scan at 120 kV exposure, non-contrast images of approximately 6 mm slice thickness can be obtained at the L4-L5 level (Fig. 9.2). These images can then be analysed using semi-automated image segmentation software. The threshold

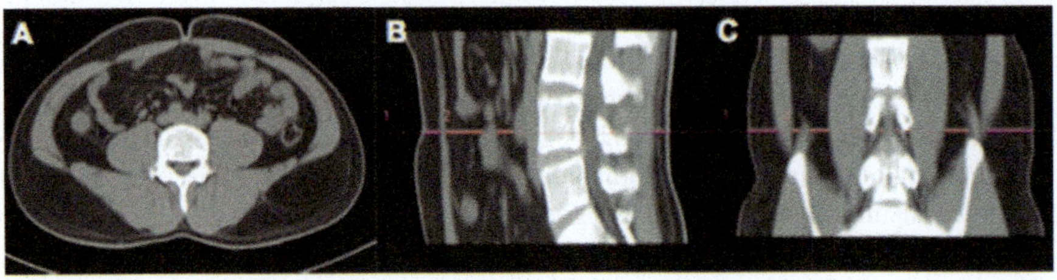

Fig. 9.1 Demonstrating method for determining abdominal fat distribution on CT scan. Axial slice obtained at the level of the umbilicus (**a**). The purple line (1) represents subcutaneous fat and the orange line (2) represents visceral fat in lateral (**b**) and antero-posterior (**c**) to pogram views

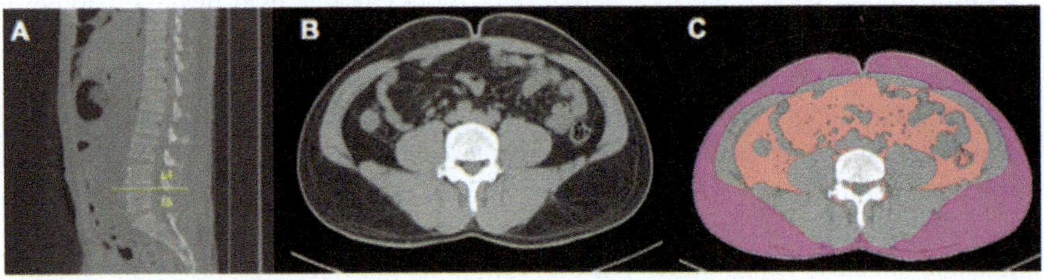

Fig. 9.2 Showing cross-sectional Computed Tomography Scan. (**a**) Yellow line showing the measurement of L4-L5 level in sagittal plane in bone window setting. (**b**) Axial slice demonstrating method for determining abdominal fat distribution. (**c**) The purple region outlines subcutaneous fat while orange region represent visceral fat compartment

value is fixed between −150 and −50 Hounsfield Units which is the range of adipose tissue on CT as previously mentioned. Adipose tissue is divided into visceral adipose tissue (VAT) and subcutaneous adipose tissue (SAT). VAT is defined as intra-abdominal fat confined within the rectus abdominis, transverse abdominis, quadratus lumborum and psoas muscles. The fat between the aforementioned muscles and subcutaneous tissue is defined as SAT.

9.4 Magnetic Resonance Imaging

When compared with CT scan, MR is a favourable as a non-radiation-exposing alternative method for determining abdominal fat. MR is an effective and precise imaging technique for computing visceral and subcutaneous adipose tissue. However, MR is expensive and less available than a CT scan.

Several MR imaging sequences can be used for imaging abdominal adipose tissue. All these techniques have advantages and disadvantages. The sequences include spin-echo T1- weighted imaging, inversion recovery sequences, Dixon techniques and water-saturation imaging. In all of these sequences, the subcutaneous and visceral adipose tissue has high signal intensity and can be easily differentiated from other tissue types.

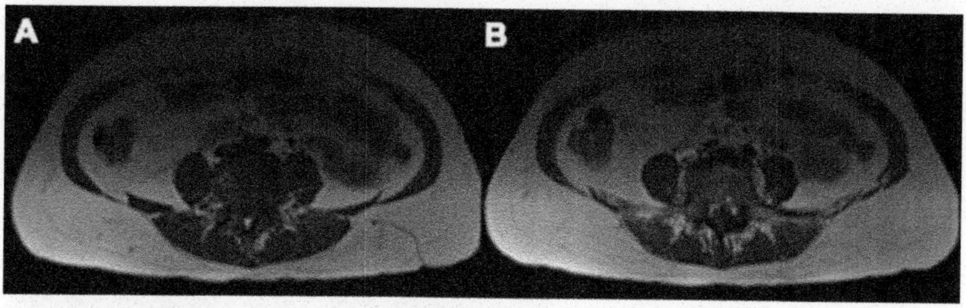

Fig. 9.3 MR imaging assessment of subcutaneous and visceral fat in T1W (**a**) and T2W (**b**). A transverse section at the level of the L5 vertebra. Fat signal is of high intensity

As whole-body MR scanning takes a long time and high costs, single-slice acquisitions are taken to represent subcutaneous and visceral body fat. The adipose tissue compartments are usually assessed on transverse sections acquired at approximately the level of the L5 vertebral body (Fig. 9.3). The fat compartments are then semi-automatically quantified using a software package.

9.5 Ultrasound

Ultrasound has been used for body composition assessment for the past 5 decades. However, despite the fact that the first paper published on the measurement of adipose tissue is as far back as the mid-1960s, it is a forgotten technology and underutilized by clinicians and researchers. Researchers proved that it is a safe, reliable and reproducible technique to determine subcutaneous and visceral fat. It is increasingly being used to differentiate subcutaneous and visceral fat (Bazzocchi et al. 2011).

The basic principle of medical ultrasound is based on the use of high-frequency sound. About 20–20,000 Hz is able to detect by humans. Ultrasound operates at a frequency of around 2–15 MHZ or sometimes higher. The ultrasound transducer varies with regard to frequency and mode. The higher the frequency causes the better the resolution, however, there is a decrease in penetration. An amplitude-mode or A-mode depends on a narrow beam to scan tissue and produces a spike on the graph. The brightness modulation or B-mode uses a linear array and produces two-dimensional images by combining A-mode signals from different directions.

The beam of the ultrasound originates from the mechanical oscillation of numerous crystals that are excited by electrical pulses in the transducer known as the piezoelectric effect. The ultrasound beam is transmitted through the skin and to the subcutaneous fat, fat, muscle or bone. It is then reflected to the transducer as an echo. Thus, the transducer serves as both a transmitter and a receiver (Wagner 2013).

The brightness of the image on the computer screen is represented by the amplitude or strength of the echoes. No echoes appear black, weak reflections appear grey and strong reflections appear white. As this produces a grey-scale image, the

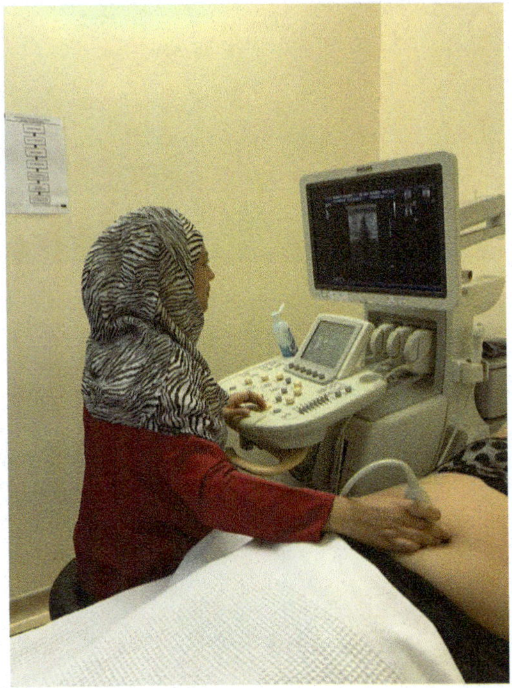

Fig. 9.4 Ultrasound scan done in transverse plane

skin-subcutaneous fat and muscle-bone interfaces appear white. The border of the fat-muscle interface is less distinct but visible.

To date, there is no standardized protocol for measuring subcutaneous fat using ultrasound. The majority of the studies have used a 5 MHz B-mode transducer. In a study by Rolfe et al. (2010) to measure visceral thickness, the transducer was placed longitudinally on the abdomen where the xiphoid line intercepted the waist circumference. The image was captured at the end of a quiet expiration to avoid tensing and distorting the abdominal cavity. Visceral thickness was then defined as the depth from the peritoneal boundary to the corpus of the lumbar vertebra. Subcutaneous abdominal fat thickness was determined at the same location but on the transverse plane (Figs. 9.4 and 9.5). The transducer is placed gently on the surface of the abdomen to avoid compression of the subcutaneous adipose fascia.

In view of its portability, ultrasound has a greater advantage over large, stationary imaging techniques such as DXA, CT and MRI. Ultrasound is safe and there is no ionizing radiation. Furthermore, when compared with other laboratory methods ultrasound is far less costly.

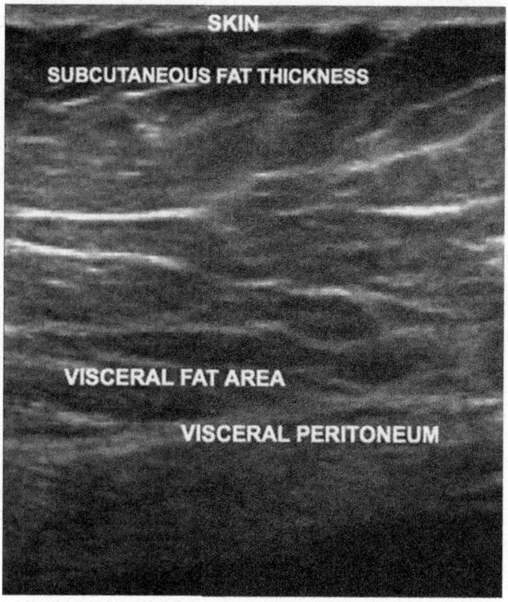

Fig. 9.5 Ultrasound image showing subcutaneous and visceral fat regions

9.6 Conclusion

Abdominal fat is distributed into two main compartments which have different metabolic characteristics: subcutaneous and visceral adipose tissue. Measuring visceral adipose tissue is extremely important as it is closely associated with metabolic syndrome eventually contributing to morbidity and mortality. The necessity for reliable and precise measurement for quantifying VAT is clear. Although there are many imaging methods available to measure subcutaneous and visceral adipose tissue, CT is probably the most commonly used method. However CT has its limitations due to exposure to ionizing radiation, and its use is especially restricted in children and adolescents. MRI provides similar accuracy however it is costly. A limitation of both CT and MRI is the requirement of expensive analysis software to quantify abdominal fat. Without a precise measurement of VAT, an index of abdominal obesity cannot be defined. Ultrasound is able to estimate subcutaneous and intra-abdominal fat however its accuracy and reproducibility are poor. Imaging methods for measuring SAT and VAT depend on factors such as availability of modality, accuracy and specificity. If CT and MRI are available, these imaging modalities probably produce the most specific and accurate methods for measuring SAT and VAT.

References

Bazzocchi A, Filonzi G, Ponti F, Sassi C, Salizzoni E, Battista G, Canini R (2011) Accuracy, reproducibility and repeatability of ultrasonography in the assessment of abdominal adiposity. Academic Radiology 18(9): 1133-1143

Cabrera MA, Gebara OC, Diament J, Nussbacher A, Rosano G, Wajngarten M (2007) Metabolic syndrome, abdominal obesity, and cardiovascular risk in elderly women. International Journal of Cardiology 114(2): 224-229

Chen HJ, Bai CH, Yeh WT, Chiu HC, Pan WH (2006) Influence of metabolic syndrome and general obesity on the risk of ischemic stroke. Stroke 37(4): 1060-1064

Gami AS, Witt BJ, Howard DE, Erwin PJ, Gami LA, Somers VK, Montori VM (2007) Metabolic syndrome and risk of incident cardiovascular events and death: a systematic review and meta-analysis of longitudinal studies. Journal of the American College of Cardiology 49(4): 403-414

Jacobs EJ, Newton CC, Wang Y, Patel AV, McCullough ML, Campbell PT, Gapstur SM (2010) Waist circumference and all-cause mortality in a large US cohort. Archives of Internal Medicine 170(15): 1293-1301

Rolfe EDL, Sleigh A, Finucane FM, Brage S, Stolk RP, Cooper C, Ong KK (2010) Ultrasound measurements of visceral and subcutaneous abdominal thickness to predict abdominal adiposity among older men and women. Obesity 18(3): 625-631

Wagner DR (2013) Ultrasound as a tool to assess body fat. Journal of Obesity 2013

Extrapolation from Clinical Trial to Practice: Current Pharmacotherapy on Obesity

10

Bapi Gorain, Hira Choudhury, Pallav Sengupta, Rohit Kumar Verma, and Manisha Pandey

Abstract

The epidemic structure of obesity has known to exacerbate or initiate comorbidities, including cardiovascular complications and diabetes. Although awareness of people through public health initiatives is struggling, it also gains attention towards pharmacotherapy, bariatric surgery and medical devices. Nowadays, the treatment of obesity with anti-obesity drugs has markedly accepted as adjunctive treatment, connecting the gap between the modification of lifestyle and bariatric surgery. Such treatments are pharmacologically effective through several mechanistic pathways, by acting through the central nervous system or on the gastrointestinal system to induce weight reduction with tolerable side effects. This chapter of the book focuses on the recent available pharmacotherapeutic agents in terms of safety and efficacy which have been approved by FDA to reduce the weight of obese patients followed by maintaining their reduced body weight.

B. Gorain (✉)
School of Pharmacy, Faculty of Health and Medical Sciences, Taylor's University, Subang Jaya, Selangor, Malaysia

H. Choudhury · M. Pandey
Department of Pharmaceutical Technology, School of Pharmacy, International Medical University, Kuala Lumpur, Malaysia

P. Sengupta
Department of Physiology, Faculty of Medicine, Bioscience and Nursing, MAHSA University, Kuala Lumpur, Malaysia

R. K. Verma
Department of Pharmacy Practice, School of Pharmacy, International Medical University, Kuala Lumpur, Malaysia

M. K. Kutty, A. Elengoe (eds.), *Obesity and its Impact on Health*,
https://doi.org/10.1007/978-981-33-6408-0_10

Keywords

Obesity · Risk · Weight loss · Pharmacotherapy · Clinical research

10.1 Introduction

Obesity is a chronic, complicated and heterogeneous disease which affects the entire body organs of patients with a high body mass index (BMI > 25 kg/m^2)(Srivastava and Apovian 2018). The occurrence of obesity is increasing at a constant rate, and it has been measured that accounting of obese patients from 1980 (BMI $\geq$ 25 kg/m^2) has been doubled in 73 countries, the consequence of which reflected by approximately four million deaths worldwide apart from the incidence of comorbid cardiovascular complications (GBD 2015 Obesity Collaborators 2017). Alternatively in 2008, more than 200 million men and approximately 300 million women were recorded obese globally (BMI $\geq$ 30 kg/m^2) (Manning et al. 2014). Therefore, the incidence of obesity has turned to the form of epidemic in most western countries, however, remaining countries are also documented with a prevalence of 30% (Flegal et al. 2012). Public awareness programmes against obesity are continuously offering initiatives, but failures on the effective goal have pointed towards alternatives to the preventive strategies (Manning et al. 2014).

Apart from improving healthy food habits for a prolonged period and continuous physical activities, a multifaceted approach to control obesity for most of the obese patients includes pharmacotherapies. Such pharmacotherapies help in reducing and maintenance of body weight. Based on the underlying safety issues of the previous therapies, current therapies are more acceptable to the users. This chapter has been focussed on currently approved pharmacotherapies for the treatment and management of obesity with an acceptable efficacy profile.

10.1.1 Need for Therapy in the Management of Obesity

It has well recognized that admiration of healthy lifestyle including modified food habits, daily physical exercise with lifestyle intervention of psychological features are the cornerstone of obesity management and can effectively reduce the excess body weight and maintain it for the short or medium obese conditions (Loveman et al. 2011; Manning et al. 2014). However, very-low-calorie diets or additional intensive meal replacements help in the management of severely obese patients (Lean et al. 2013). Moreover, management of obesity with such implements could reverse the initial condition of the patients or even worsen with time. However, bariatric surgery has become popular nowadays for the severe (BMI $\geq$ 40 kg/m^2) or even complex obese conditions (BMI $\geq$ 35 kg/m^{2}with coexistence of obesity-related co-morbidity (ies)). This has become popular due to its effectiveness within a short period to achieve a great loss of weight which subsequently recover the patients from obesity-related co-morbidities (Kashyap et al. 2013; Sjöström et al. 2012). Bariatric surgery also has been shown to help in life expectancy by reducing of causes of mortality associated with obesity (Sjöström et al. 2007). Such surgeries help in a substantial, sustained and significant reduction of the body weight, whereas the associated mortality rate is also very low (0.08–0.35%). Besides its effectiveness, although low, associated complications of the surgery range from 10% to 17%, where the reoperation rate of the patients is 7% (Chang et al. 2014) and sometimes resulted in regain of weight (Karmali et al. 2013). Such complications of bariatric surgery lead to the path to obtain an alternative therapeutic strategy with a more effective and acceptable approach. Therefore, the researchers have considered the pharmacotherapeutic treatments as an effective alternative approach against the surgeries when it must be combined with the controlled diet, physical exercise and behavioural interventions (Manning et al. 2014). Breakthrough in recent research to identify the disease homeostasis and underlying pathological mechanism for the overweight tendency and maintenance of obesity, several potential targets have been targeted to achieve success in pharmacotherapy (Guyenet and Schwartz 2012). Furthermore, the novel therapeutic agents for obesity therapy have been focused on long-term therapy and when a patient responded significantly towards this medication (generally defined as $\geq$5% loss of weight over a treatment period of 12 weeks), continued for an undefined time (Wharton 2016). Consequently, combination strategies of pharmacotherapy and lifestyle intervention have shown to be a greater control and maintenance concept of weight reduction, where modification of lifestyle reduces the weight of the obese patients whereas treatment thereafter aids in maintenance in the reduced bodyweight (Srivastava and Apovian 2018).

10.2 Challenges in the Obesity Treatment

10.2.1 Cause and Risk

A chronic surplus of calorie consumption compared to calorie expenditure would create a state of energy imbalance which subsequently results in excessive accumulation of triglycerides in adipose tissue and thus obesity.

Parental obesity has been well established to be the strongest risk factor for childhood and adolescent obesity, with the risk being further elevated if both parents are obese, and therefore reinforces the genetic dimension of obesity (Reilly et al. 2005; Whitaker 2004). Undeniably, genetics do play a part, but the small difference in BMI, combined with the substantial increase in the prevalence of obesity over the last few decades in developed and developing nations; suggests that other risk factors beyond genetics such as diet and exercise are also implicated in obesity risk. Diet and physical activity affect the balance between caloric intake and caloric expenditure for the regulation of weight (Ghee 2016). Undoubtedly, evidence from clinical trials have almost consistently pointed to better weight management with caloric restrictions, there are however a few barriers in controlling calorie intake, with the most common reasons cited among obese patients were "attend meetings often" (60.0%) and "difficult when eating out" (57.9%) (Lim et al. 2014; Wadden et al. 2012).

The literature is flooded with information on the roles of diet quality besides overall caloric intake. For example, a study conducted among health professionals residing in the United States reported a robust positive association between averaged 4-year weight gain with the intake of potatoes, potato chips, sugar-sweetened beverages, and red meats, but an inverse association with the consumption of whole grains, fruits, vegetables, nuts, and yogurt (Mozaffarian et al. 2011). Moreover, a concomitant increase in the consumption of added sugar in sugar-sweetened beverages, with the incidence of obesity, has been observed (Lim et al. 2014; Malik et al. 2010). The obesity epidemic is also complicated by sedentary behaviours beyond diet (sleep, physical activity, screen time), which perse is independently associated with weight change, and when combined with dietary factors, have cumulative and additive effects on weight regulation (Ibrahim et al. 2014). Review of findings from randomized trials and observational studies uniformly reporting that, in general, 150–250 min per week of moderately intense activity is warranted to prevent weight gain, or aid in weight loss, when coupled with dietary modification (Donnelly et al. 2009). Other than physical activity, leisure-time activities which are not truly restful behaviors, such as getting less than 6 h or more than 8 h of sleep in adults and adolescents or less than 10–11 h of sleep in children, and television viewing time, have also been linked with weight gain (Boone et al. 2007; Hu et al. 2003; Patel and Hu 2008).

Environmental factor has a significant impact upon the extent to which the physiological homeostatic mechanisms of our body act to maintain body weight in the healthy range. A few decades ago when the global environment was the one where extensive physical activity was required to secure food and shelter as well as for mobility, with a limited food choice available, the physiological homeostatic mechanisms of our body act to encourage sufficient nutrient absorption to avoid loss of body weight. However, alongside impressive demographic and economic development, the global melieu has gradually morphed to the one which is conducive to weight gain, or "obesogenic" state, where extensive physical activity is no longer indicated in our daily life, with an abundance of largely inexpensive food choice served in bigger portion sizes. In these settings, the physiological maintenance of body weight becomes inadequate in opposing the positive energy balance, and the development of obesity appears to be an adaptation to the environmental changes (Manning et al. 2014; Swinburn et al. 2011; Wadden et al. 2012).

10.2.2 Expectations and Goals

The first line of treatment for obese patients having BMI more than 30 is anti-obesity drugs which sometimes can't meet the expectation of doctors as well as the patients. The clinician always attempts to choose the ideal drug to reduce visceral fat as well as to manage metabolic perturbance associated with the pathogenesis of obesity to enhance the quality of life. Additionally, ant obesity drugs are available at an affordable price with less adverse effects for long term use. Bariatric surgery is recommended in patients with BMI more than or equal to 40 kg/m^2 or $\geq$ 35 kg/m^2

with at least one obesity-related comorbidity. However, pharmacotherapy is recommended in patients with BMI $\geq$ 30 kg/m^2 and BMI $\geq$ 27 kg/m^2 with at least one comorbidity, provided dietary control and physical exercise cannot maintain weight-loss goals. The selection of drugs for weight reduction should be in conjunction with comorbidity associated with the patients. The interactions of the obesity pharmacotherapy with both the comorbid conditions and concomitant medications should be considered (Golden 2017; Manning et al. 2014; Swinburn et al. 2011; Wadden et al. 2012).

Insufficient comprehension of aetio-pathogenesis of obesity has rendered further research on targeted therapies for obesity (O'Rahilly and Farooqi 2008). Leptin replacement is one of the examples of obesity-targeted therapy that is exceedingly rare congenital leptin deficiency (Paz-Filho et al. 2011). As referred to causes and hazard, genetic factor, ease of food accessibility and daily life routine is the source of imbalanced energy intake and disbursement which culminates in surplus body fat (Speakman et al. 2011; Swinburn et al. 2011). Therefore, prior to pharmacotherapy, increased physical activity, dietary changes to achieve caloric deficit and behavioral therapy are the keystone of weight-loss in patients (Garvey and Mechanick 2016; Yumuk et al. 2015). Despite this available option for management of obesity, long-term weight loss management is disappointing in the majority of patients because of compensatory mechanisms associated with weight loss (Jensen et al. 2014; Yumuk et al. 2015).

10.3 Current Achievement in the Treatment of Obesity

Stringent guidelines of FDA on screening of anti-obesity agents became effective since the mid-1990s where the therapy should statistically induce a substantial placebo-adjusted weight loss of more than 5% at 1 year or 35% of patients should achieve more than 5% weight loss (Department of Health and Human Services 2007). Additionally, the therapy should adduce the evidence in the improvement of metabolic biomarkers, such as control in diabetes, hypertension, or hyperlipidaemia (Department of Health and Human Services 2007; Srivastava and Apovian 2018). Thus, current anti-obesity agents, such as orlistat, phentermine, phentermine-topiramate, lorcasein, liraglutide, naltrexone-bupropion are the adjunct treatment for obese patients with aforementioned comorbidities with BMI $\geq$30 kg/m^2, to be combined with lifestyle interventions, including increased physical exercise and low-calorie diet. A descriptive discussion of the drugs has been provided in the following section with a clear structure of the drugs profile in Table 10.1.

10.3.1 Orlistat

Orlistat or tetrahydrolipstatin has been approved by the FDA for obesity pharmacotherapy in 1999 in conjunction with decreased calorie intake. This drug is also

Table 10.1 Currently available FDA approved obesity pharmacotherapies

Name of the approved medication	Dosage and administration	Mechanism of anti-obesity action	Adverse effects	Interactions	Warning	Contraindication
Orlistat	60 mg (OTC) or 120 mg; 120 mg orally TID dailywithin 1 h of fat-containing meal	It is a pancreatic and gastric lipase inhibitor, inhibits fat to get absorbed	Fatty/oily stools, bloating, flatulence, cramps, fecal incontinence, and diarrhoea	As the drug acts locally in the GIT, it interacts with number of drugs and interfere their absorption, bioavailability and ultimately their pharmacological response, for e.g., warfarin, cyclosporine, amiodarone, levothyroxine, and antiepileptic drugs	Malabsorption of oil-soluble vitamins (A, D, E and K)	Chronic malabsorption condition, cholestasis,breast feeding and pregnancy. Some medications (cyclosporine, fat-soluble vitamin supplements and analogues, levothyroxine, warfarin, amiodarone, antiepileptic drugs
Phentermine	A 8 mg oral dose before meals orally TID daily, OR, 15–37.5 mg just before breakfast/ 1–2 h after breakfast	It is a sympathomimetic amine acting centrally	Headaches, hypertension, increasedHR, anxiety,tremor, constipation, dry mouth, irritability, insomnia, dizziness, palpitations, change in libido	Prescription of this drug is restricted with other appetite-suppressant drugs, monoamine oxidase inhibitors, tricyclic antidepressants, phenothiazines, serotonin–norepinephrine reuptake inhibitor, selective serotonin reuptake inhibitors,	Serious regurgitant cardiac valvular disease and primary pulmonary hypertension can be observed if coadministered with fenfluramine or dexfenfluramine	Cardiovascular disorder, anxiety disorder, uncontrolled hypertension, hyperthyroidism, agitated states, glaucoma or MAOI users, and pregnant and breast feeding women

				other CNS stimulants, or drugs that induce seizure and dizziness		
Lorcaserin	Either 10 mg oral dose BID daily; or extended release version oral single dose of 20 mg daily	This agent is a 5-HT2c receptor agonist, acting centrally	Headache, nausea, dry mouth, dizziness, fatigue, constipation are common in normal obese patients. Fatigue, back pain, headache, upper respiratory tract-like symptoms, like cough, and hypoglycaemia in T2DM obese patients	Prescription of this drug is restricted withsome medications (SNRIs, SSRIs, MAOIs,triptans, bupropion, dextromethorphan, St John's Wort)	Neuroleptic malignant syndrome or serotonin syndrome may arise when co-administered with contraindicated drugs may result in cognitive impairment, suicidal ideation and valvular heart disease	Pregnancy and breastfeeding women
Liraglutide	Daily dose of 0.6 mg with gradual weekly increaseof 0.6 mg/week until the dose of 3.0 mg is achieves; SC	This agent is a GLP1 receptor agonist, decreases hunger and increases satiety	Nausea, vomiting, diarrhoea, fatigue, constipation, headache, decreased appetite, dizziness, hypoglycaemia, dyspepsia, abdominal pain, increase in lipase levels and suicidal behaviour or ideation. Gall		Thyroid C-cell tumours, acute gallbladder disease, suicidal tendency, increase in HR, acute pancreatitis, renal impairment and hypoglycaemia	Family or personal history of medullary thyroid cancer, or multiple endocrine neoplasia type 2 (MEN2) syndrome, pregnancy, pancreatitis or to the patients hypersensitive to liraglutide

(continued)

Table 10.1 (continued)

Name of the approved medication	Dosage and administration	Mechanism of anti-obesity action	Adverse effects	Interactions	Warning	Contraindication
			bladder related adverse events are also evident			
Phentermine–topiramate	Starting dose is 3.75/23 mg for first 2 weeks, followed by double the dose (7.5/46 mg). Monthly upward titration required to achieve significant weight loss, maximum dose may be 15/92 mg: Drug can be discontinued if <3% weight loss achieved on 11.25/69 mg or < 5% weight loss on maximum after 12 weeks	Combination of the extended release products acts through sympathomimetic amine and anorectic action	Peripheral neuropathy (usually transient), dyspepsia, dysgeusia, constipation, insomnia and dry mouth	The combination of drugs interacts with MAOI, oral contraceptives, agents causing depression including alcohol, antiepileptic drugs, carbonic anhydrase inhibitors, and non-potassium sparing diuretics	Teratogenicity, metabolic acidosis, changes in memory or concentration, suicidal ideation, rise in creatinine levels, hypokalaemia, increase in HR, acute myopia or acute angle-closure glaucoma	Pregnancy and breastfeeding, poorly controlled hypertension, hyperthyroidism, glaucoma, and pre-existing treatments as depicted in interaction column
Naltrexone–bupropion	Initiate treatment with 8/90 mg for 1 week followed by weekly upward titration to reach to target dose of 32/360 mg (BID)	Combination opioid antagonist and aminoketone antidepressant	Nausea, vomiting, diarrhea, headache, constipation, dry mouth, and dizziness	Combination of drugs has shown to interact with MAOIs, drugs which are metabolized by CYP2D6, inhibitors	Suicidal ideation, decrease in seizure threshold, hepatotoxicity to naltrexone component, acute angle glaucoma,	Uncontrolled hypertension, binge-eating disorder, pregnancy, anorexia nervosa or bulimia, seizure disorder,

				or inducers of CYP2B6, drugs used to lower seizure threshold, and dopaminergic drugs	hypertension and/or increase in HR	abrupt drug or alcohol withdrawal, chronic opioid use, MAOI

effective to maintain the weight of a patient after prior weight loss. This agent acts by inhibiting the lipases (pancreatic and gastric) and prevents hydrolysis of triglyceride in the gut, thus inhibit the absorption of consumed fats up to a level of 30% from the diet (Henness and Perry 2006). As mentioned in Table 10.1 that orlistat prevents the absorption of the fat-soluble vitamins (Vitamin A, D, E and K), co-administration of multivitamin consisting of vitamin A, D, E, K and beta-carotene is highly recommended. The adverse events alluded to in Table 10.1 are mostly associated with the excretion of dietary fats in the stool. Thus a low-fat diet is always advisable for the drug users, moreover in case of ignorance very awkward and problematic side effects results.

Orlistat is one of the most studied pharmaceutical agents for the treatment of obesity. It has been evidenced in the literature with >150 clinical trial records for orlistat, either alone or combination for the management of obesity and also in the control of different aspects of the metabolic disorder (McClendon et al. 2009). A meta-analysis report of 45 clinical trials by Dombrowski and team has shown that orlistat along with behavioral interventions able to reduce overall body weight by 1.8 kg within a treatment period of 12 months, however reduction of weight was found dose-dependent. A dose of 120 mg, TID has shown better maintenance in weight reduction (2.34 kg), compared to the other lower doses (0.7 kg for 60 and 30 mg) (Dombrowski et al. 2014). Further, orlistat has also shown to lessen the incidence of T2DM (6.2% of orlistat treated patients as compared to 9.0% with placebo) consequent to a reduction of risk by 37.3% (Torgerson et al. 2004) along with the decreased level of LDL and total cholesterol irrespective of weight loss (Muls et al. 2001).

10.3.2 Phentermine

Phentermine is the centrally acting anti-obesity agent, approved in the year 1959 by the FDA to be used over the age group of 16 years in the USA. This drug is a sympathomimetic amine that acts *via* an insignificant increase of dopamine release with the CNS, which in turn suppresses the appetite of the users through pro-opiomelanocortin (POMC) neuronal activation (Hampp et al. 2013; Jordan et al. 2014; Srivastava and Apovian 2018). This drug can only be prescribed for short-term use because the patient develops drug tolerance and deleterious effects. Increasing the dose can overcome the tolerance; however, FDA restrictions suggest discontinuing its use. The addictive property of the drug restricts its further use; the drug will be withdrawn if there is any addictive sign monitored. The dosage and schedule, adverse events, interactions and contraindications of this drug have been depicted in Table 10.1.

Clinically, the role of phentermine has been evidenced in many clinical studies. A 12 weeks of phentermine treatment achieved ≥5% weight reduction in 95.8% of patients, whereas 20.8% in the placebo group (Kang et al. 2010). In another recent trial of the monotherapy of the drug compared with combinations, a 5.1% weight loss had been reported at 28 weeks with monotherapy (Aronne et al. 2013). As discussed earlier that the drug can be appropriately prescribed for the short term, the off-label long term application of the drug has also been tested by The Endocrine Society Pharmacological Management of Obesity, however absence of clinical risk factors with targeted role paves avenues for further studies with the selected dose of 7.5–15 mg (Apovian et al. 2015).

10.3.3 Lorcaserin

Although the approval process of lorcaserin encountered various hurdles, FDA approves this drug for the management of overweight (BMI $\geq$27 kg/m^2) who have any one of the weight-related comorbid condition (hypertension, diabetes, hyperlipidemia, or sleep) or obese adults (BMI $\geq$30 kg/m^2) in 2012 in combination with lifestyle interventions, such as augmented physical exercise and intake of a low-calorie diet (Morrato and Allison 2012; Taylor et al. 2013). Serotonin is considered a modulator of food intake and appetite (Greco et al. 2017; Lam et al. 2010). Lorcaserin is a serotonin receptor agonist, working through 5-HT$_{2c}$ receptors in the central nervous system, particularly in the hypothalamus. Agonistic action of the drug acts on anorexigenic POMC neurons and activates the neurons, thereby suppresses appetite (Rogge and Gautam 2017). Like phentermine, lorcaserin is also a schedule-IV agent according to the Controlled Substances Act and psychic dependence can be developed if abused with a supratherapeutic dose (40–60 mg). Further, lorcaserin is considered to be safe to prescribe to patients with cardiovascular complications, in conditions phentermine/sympathomimetics should not be prescribed. Lower potential to increase HR and BP facilitates its use in obese patients with cardiac complications (Smith et al. 2010). A recent trial on lorcaserin revealed its potential of safety in pre-existing FDA-defined valvulopathy patients following therapy of 52 weeks (Weissman et al. 2017).

Three major clinical trials of lorcaserin have been conducted for the management of overweight and obesity (BLOSSOM and BLOOM) and another study for the management of obesity and type 2 diabetes mellitus (T2DM) (BLOOM-DM) (Srivastava and Apovian 2018). Included subjects in the first two trials received lorcaserin 10 mg BID, where 47.1% patients achieved $\geq$5% loss of body weight following treatment for 1 year, whereas 22.6% achieved with placebo; patients in both the groups were receiving lifestyle interventions (Fidler et al. 2011; Smith et al. 2010). In the third study in T2DM obese patients, BID lorcaserin 10 mg treatment revealed 37.5% patients with $\geq$5% loss of body weight, whereas the placebo group achieved 16.1% following treatment and lifestyle intervention for 1 year (O'Neil et al. 2012).

10.3.4 Liraglutide

Glucagon-like peptide-1 (GLP1), liraglutide, has been approved for its effective control of hyperglycemia in T2DM patients in 2010. However, later its use and dosage have been modified for the management of overweight patients with a BMI $\geq$ 27 kg/m^2 adjunct to a controlled diet and physical exercise (Mehta et al. 2017). GLP1 is an incretin receptor available centrally as well as peripherally and plays an important role in the signaling cascade for glucose-induced insulin secretion (Madhu et al. 2018). Thus, *via* the central and peripheral pathways, liraglutide alters glucose balance, intake of food and satiety. Action on the hypothalamus has been presumed to reduce hunger and subsequently results in increased satiety. It is the only injectable anti-obese pharmacotherapy for subcutaneous administration as a suspension formulation (Secher et al. 2014; Srivastava and Apovian 2018). Dose titration for the drug from 0.6 mg to 3.0 mg/day is conducted to reduce associated GI-related side effects. Further, if the drug shows any signs of pancreatitis, the drug will be withdrawn and will not be continued (Golden 2017).

Treatment of liraglutida (3.0 mg) along with lifestyle modification over 56 weeks demonstrated loss of $\geq$5% body weight in 44–62% of the obese participants, whereas it is 16–34% in the placebo group (Golden 2017). Use of liraglutide 3.0 mg caused an increase of HR (2 beats/min) in patients, relative to the placebo, however this increase in HR is not dose-dependent, and this applies to the dosage of 1.8 mg or 3.0 mg (Golden 2017). Additionally, treatment with liraglutide may be beneficial for obese patients with cardiovascular complications, as the LEADER trial reported decreased portability in the liraglutide treatment group (1.8 mg/daily) compared to placebo (Marso et al. 2016). Further, the suicidal tendency (0.2%) of the drug has also been monitored during clinical research, therefore monitoring of depression and suicidal ideation is essential for the patients on liraglutide. As mentioned in the warning section of this drug in Table 10.1, this drug has also shown to produce thyroid C-cell tumors in animal models (Golden 2017).

10.3.5 Phentermine/Topiramate

The description of phentermine has already been provided earlier. In 2012, the extended-release combination of phentermine and topiramate has been approved for weight management for overweight persons. Pharmacologically topiramate is an anti-epileptic agent, which is also prescribed for migraines. The underlying mechanism of its anorexic role remains unveiled hitherto. But it has been postulated that stimulation of neurotransmitters in the CNS, including inhibitory effects on voltage-gated Na^+ channels, carbonic anhydrase and glutamate receptors, and potentiation of GABA activity may participate in the anorexic role of topiramate (Cosentino et al. 2013). The amalgamation of these two components provides a synergistic effect on obese patients for the reduction and management of overweight than the monotherapies. Concomitantly, the use of phentermine monotherapy is restricted because of its abuse potential; however, the combination product is also kept under schedule-IV according to the Controlled Substance Act. The adverse events of the combination have been depicted in Table 10.1; in case the patients experience paraesthesia or tingling sensation, the dose will be reduced in order to resolve the adverse events. If, on the other hand, it is ignored, may lead to peripheral neuropathy. Inhibition of carbonic anhydrase by topiramate may also aggravate the formation of kidney stones by reducing citrate excretion from the kidneys and by increasing urinary pH (Golden 2017). The antiepileptic agent may also induce suicidal ideation, whereas the combination may shift the mood towards anxiety or depression, hence proper monitoring of the consumers is highly recommended (Halpern and Mancini 2017; Shin and Gadde 2013). Moreover, the combination has exhibited interaction with MAOI, where it induces hypertensive crisis with resultant hypokalemia due to the non-potassium sparing diuretics, whereas cognitive issues or dizziness are common with CNS depressants. Further, monitoring of serum creatinine, glucose and electrolytes are mandatory if the combination is treated with anti-hyperglycemics, as co-administration may result in metabolic acidosis and hypoglycemia (Cosentino et al. 2013).

Two important double-blind clinical trials have been conducted to assess the efficacy of the combination over 52 weeks with a 4-week titration period EQUIP (Controlled-Release Phentermine/Topiramate in Severely Obese Adults) and CONQUER (Controlled-Release Phentermine plus Topiramate Combination in Overweight and Obese Adults). One of these studies was extended further for 2 years with an additional 52 weeks treatment to the completers of CONQUER study subjects, SEQUEL (2 year Sustained Weight Loss and Metabolic Benefits with Controlled-release Phentermine/Topiramate in Obese and Overweight Adults). All studies were performed in conjunction with intensive lifestyle modification including a well-balanced and reduced-calorie diet coupled with physical exercise (Lonneman et al. 2013). Clinical studies have revealed enormous success to use the combination as compared to the placebo in managing overweight persons by reducing body weight and increasing physical activity. A greater proportion of obese patients achieved a weight loss of $\geq$5% with the active treatment group (62% with 7.5/46 mg and 67% with 15/92 mg) (Cosentino et al. 2013).

10.3.6 Naltrexone–Bupropion

A proprietary formulation of extended-release of combined naltrexone and bupropion has been approved by US FDA in 2014 for long-term treatment and management of weight in obese patients (Yanovski and Yanovski 2014b). Naltrexone is an opioid antagonist, whereas bupropion is an antidepressant. Animal studies suggest the action of the combination to the appetite-regulating centre at the hypothalamus and in the mesolimbic dopamine circuit. The combination is synergistically acting on POMC centre to produce a collective effect in reducing hunger and satiety (Wang et al. 2014). Individually, naltrexone is an opioid antagonist, causes opioid receptor-mediated POMC auto-inhibition and bupropion particularly inhibits neuronal reuptake of neurotransmitters, such as dopamine and noradrenaline. As bupropion is an anti-depressant, and the drug combination is considered under Control Substance Act, the use of the combination should be monitored for depression or suicidal ideation during the early stages of treatment (Plodkowski et al. 2009).

Four clinical trials on this combination have been conducted for 56 weeks versus placebo, Contrave Obesity Research I (COR-I), COR-II, COR-Behavioural Modification (COR-BMOD) and COR-Diabetes in either obese patients with 30–45 kg/m^2 BMI or 27–29.9 kg/m^2 BMI with evidence of hyperglycemia with or without dyslipidaemia and/or hypertension (Hong et al. 2016; Sherman et al. 2016). COR-I and COR-II clinical trial results showed a 36–42% achievement in reduction of ≥5% weight with the combination of naltrexone–bupropion with lifestyle intervention, whereas it was only 17–18% in the placebo group with lifestyle modification (Hong et al. 2016; Sherman et al. 2016).

10.4 Clinical Aspects

Obesity deviates the body from a homeostatic state and serves as the underlying cause of various comorbidities, such as hypertension, cardiometabolic disorders and T2DM. Weight-loss treatment via pharmacological approach has consistently been a promising research area, while non-pharmacological methods have also gained immense popularity. As early as in the 1950s and 1960s, the development of centrally acting sympathomimetics such as phentermine and diethylpropion was initiated (Colman 2005; Wilding 2007). Fenfluramine and dexfenfluramine (5-HTreleasing agents) were later implemented as key drugs in obesity treatment in the 1970s and 1980s. The use of these 5-HT releasing agents came with side-effects, mainly causing pulmonary hypertension. Moreover, the combination treatment with fenfluramine and phentermine showed greater efficacy than monotherapy, in the early 1990s (Weintraub et al. 1992), while this too can cause cardiac valvulopathy (Connolly et al. 1997), and consequently lead to the withdrawal of therapy. Therefore, the FDA approved these sympathomimetic agents for short-term use only (up to 12 weeks). Phentermine was the most commonly prescribed obesity drug in the United States from 2008 to 2011 (Hampp et al. 2013). Placebo-controlled phentermine (30 mg/day) trial was performed in 108 women for the duration of 36 weeks. Women who took the medication continuously had reported an average of 12.2 kg body weight loss, while the ones who underwent intermittent medication had reported 13 kg average body weight loss and the placebo group reported a 4.8 kg average loss of body weight. Sympathomimetic drugs, other than phentermine, are less commonly prescribed. In the late 1990s, a dual monoamine (noradrenaline and serotonin) reuptake inhibitor, sibutramine, was introduced to clinical practice (Luque and Rey 2002; McNeely and Goa 1998), which resulted in moderate weight loss by increasing energy expenditure and decreasing energy intake. But the drug was soon suspended from the market after identification of a significant cardiovascular risk during its post-marketing clinical trial (James et al. 2010). In the mid-1990s, A CB1 receptor antagonist, Rimonabant was developed to curb appetite and reduce energy intake, as CB1 cannabinoid has a well-established role in appetite (Kirkham 2009). In four major clinical trials, this drug showed a consistent loss of 4–5 kg of body weight (Després et al. 2005; Pi-Synyer et al. 2006; Scheen et al. 2006; Van Gaal et al. 2005). As an anti-obesity agent, Rimonabant was approved in 2006 in Europe. It was eventually suspended in 2008 because of allegations of severe mental disorders that included anxiety, depression and even suicide. The development programmes included other CB1, including taranabant, surinabant, otenaba and ibipinabant, which led to the closure of the programme (Plieth 2008).

For the treatment of obesity, orlistat was approved by the FDA in 1998, as a gastrointestinal lipase inhibitor. Orlistat decreases intestinal fat absorption by ~30% (Borgström 1988). Double strengths of orlistat, 60 mg and 120 mg are available in the market. Orlistat of 120 mg dosage was approved only for adults and teenagers. About 35–73% of the patients achieved 5% loss in weight in a year in a clinical trial using orlistat (120 mg), while 10% loss of body weight was achieved by 14–41% of the patients. The trials continued for the second year including a weight-management diet, which showed that patients taking 60 mg of orlistat lost around 2.5 kg more than the placebo group that corresponded to about 2.5% of the original body weight; and patients taking 120 mg orlistat lost about 3.3 kg more than the placebo group, corresponding to almost 3.3% of their initial body weight (Davidson et al. 1999; Finer et al. 2000; Rössner et al. 2000; Sjostrom et al. 1998). Moreover, a meta-analysis of 3951 studies showed that orlistat can significantly decrease BMI and body weight in obese patients with the polycystic ovarian syndrome (PCOS), as well as reduction in Homeostatic Model Assessment of Insulin Resistance (HOMA-IR), testosterone and insulin (Graff et al. 2016). Another study on 52, euglycemic and euthyroid, obese female patients, showed that orlistat treatment in combination with hypocaloric diet effectively reduced body-weight as well as increased the serum concentrations of the gut-derived hormones, ghrelin and obestatin (Pamuk et al. 2018). Orlistat (120 mg) treatment for one-year reportedly is associated with substantial improvement in cardiovascular health including a reduction in systolic and diastolic pressure, fasting glucose, and cholesterol (Johansson et al. 2010; Zhou et al. 2012). This can be attributed either to the loss of weight or other related weight loss-independent factors (Erdmann et al. 2004). The limitation of Orlistat use is attributed to its primary mode of action that adversely affects the gastrointestinal functions. However, co-intake of fibre-containing food supplements may minimize these adverse incidents (Cavaliere et al. 2001). Orlistat is the only anti-obesity drug approved for long-term clinical use in Europe between 2011 and 2012 (Rodgers et al. 2012).

Another agent that underwent several clinical trials before it got FDA approval in 2012; is Lorcaserin. It was first rejected in 2010 because of tumour growth problems reported by its use in preclinical studies, however, it showed promising results during the two major clinical trials in nondiabetic patients and the third smaller trial on type2 diabetes trial (Fidler et al. 2011; O'Neil et al. 2012; Rodgers et al. 2012; Smith et al. 2010). In these trials, lorcaserin (10 mg) were given to the participants along with regulated nutrition and exercise. The average weight loss with lorcaserin treatment for 2 years was 5.6 kg, while it was only 2.4 kg in placebo participants (Chan et al. 2013). A substantial decrease in total cholesterol, LDL, triglycerides, and blood pressure was demonstrated in addition to body weight loss in non-diabetic participants. Moreover, Lorcaserin in patients with diabetes resulted in a reduction in hemoglobin A1c (HbA1c) besides the loss in body weight (O'Neil et al. 2012). The side effects of Lorcaserin included dizziness, fatigue, headache, and nausea. In addition, incidences of hypertension and valvulopathy have been reported with Lorcaserin and even though the results were not statistically significant, the FDA recommended that long-term cardiovascular effects be evaluated in a post-approval trial (Colman et al. 2012). A pilot study with lorcaserin and phentermine combinations was performed in 2015. In contrast with lorcaserin alone, the addition of phentermine almost doubled the weight loss (Smith et al. 2015).

Liraglutide, glucagon-like peptide 1 (GLP-1R) agonists, demonstrated a weight loss of up to 10 kg in patients in 2 years (Astrup et al. 2009, 2012). A 20-weeks duration trial (double-blind, placebo-controlled) was conducted where oral orlistat (120 mg) was given thrice per day or liraglutide (1.2, 1.8, 2.4, or 3.0 mg) was administered subcutaneously once per day. Treatment with liraglutide doses resulted in a mean body weight loss of 4.8 kg, 5.5 kg, 6.3 kg, and 7.2 kg, while orlistat showed 4.1 kg weight loss and the placebo group showed only 2.8 kg weight loss. Liraglutide also demonstrated the lowering of blood pressure in all doses. It also resulted in a decreased prevalence of prediabetes with a dose of 1.8–3.0 mg per day. It showed some adverse impacts including nausea and vomiting which rarely lead to discontinuation of the treatment (Astrup et al. 2009).

Rosuvastatin is an HMG-CoA reductase inhibitor with unique pharmacokinetic characteristics. It exhibits the high potential of low-density lipoprotein cholesterol (LDL-C) lowering and with its low penetration in extrahepatic tissue and low CYP3A4 interactions (Graff et al. 2016). Rosuvastatin stands with great advantages to be used for weight loss treatment. Rosuvastatin (10 mg once daily) showed an improvement mainly in visceral adiposity as evidenced by a 26 week-clinical trial on 54 middle-aged (40–65) obese participants. A meta-analysis, aimed to put forth the impact of "exercise and pharmacological interventions on visceral adiposity", analyzed the effects of rosuvastatin, rimonabant, liraglutide, metformin, gemfibrozil, orlistat, ezetimibe and empagliflozin. The study showed higher effectiveness of the pharmacological interventions including that of rosuvastatin in the reduction of visceral adiposity owing to their regulated approach (Rao et al. 2019). Rosuvastatin also showed significant improvement in vaspin serum levels in obese patients with the acute coronary syndrome (Al-Kuraishy et al. 2018). Vaspin is considered an endocrine factor, adipose tissue-derived serine-protease inhibitor which plays an essential role in the prevention of atherosclerosis and, also reportedly inversely correlates with blood pressure in obese patients (Al-Kuraishy et al. 2018). The effect of Rosuvastatin in increasing vaspin levels suggests the cardio-ameliorative properties of this drug.

The antiepileptic drug, zonisamide (400 mg), is reportedly effective in body weight reduction in obese adults in combination with well-regulated diet and lifestyle modifications (Gadde et al. 2012). Zonisamide, at a dose of about 200 mg showed weight loss in even children and adolescents who were prescribed psychiatric medications (Shapiro et al. 2016). Earlier, several studies have shown the efficacy of zonisamide in weight loss, also reported high incidences of adverse psychological effects and gastrointestinal problems (Gadde et al. 2012). Since zonisamide is an antiepileptic drug, if the drug is used for weight-loss and sudden discontinuation may induce seizures. Thus, it is suggested that zonisamide treatment cessation should follow a gradual titration-off (Stanford 2019).

Previously, only a few anti-obesity drugs were available, but recently this landscape has changed with new generation drugs as well as combination therapies are showing more promising results. Naltrexone/bupropion is a new-generation combination therapy for obesity where naltrexone (32 mg) and bupropion (360 mg) have been shown to reduce hunger by acting upon the hypothalamic hunger center mainly via modulation of the mesolimbic reward pathway. This combination treatment, in addition to an adjunct to regulated diet and exercise, is highly effective in chronic weight management in obese adults as well as in mitigating obesity-associated comorbidities (Barrea et al. 2020).

Phentermine plus topiramate-extended release (ER) treatment, received FDA approval in 2012, the show is a potent option for chronic weight- management in obese and overweight patients when combined with adjusted diet and exercise (Allison et al. 2012; Gadde et al. 2011; Kelly et al. 2013). The phentermine/topiramate-ER (7.5/46 mg) resulted in a weight loss of around 8.1 kg (7.8%) in a year and with a higher dose of phentermine/topiramate-ER (15/92 mg), the weight loss was 10.2 kg (9.8%), while there was only 1.4 kg (1.2%) weight loss in the placebo group (Saunders et al. 2016). Moreover, both the prescribed and the higher doses of the phentermine/topiramate-ER were associated with reductions in comorbidities, such as cardiovascular risk factors and the prevalence of type-2-diabetes. (Davidson et al. 2013). However, a minor rise in resting heart rate was observed with a high dose of this combination treatment (56.1%) vs placebo (42.1%). This observation raised questions about the future long-term implication of this combination therapy upon cardiovascular functions (Yanovski and Yanovski 2014a).

The efficacy of combination treatment with canagliflozine (300 mg) and phentermine (15 mg) in nondiabetic obese and overweight subjects was studied. For this randomized, double-blind and placebo-controlled study (phase-2b), 335 participants were enrolled for 26-weeks. The participants received oral doses of canagliflozin and phentermine once per day. Results showed significant weight loss by 5% in the treated group as compared to the placebo treatment group with a reduction in systolic blood pressure was decreased (Hollander et al. 2017).

10.5 Conclusion

Obesity is a major global concern that is presented with numerous comorbidities. Although diet, exercise and lifestyle management are the mainstays in the management of obesity, it is difficult to sustain a regulated lifestyle in the long-term resulting in regaining body weight. Once the non-pharmacological interventions fail, long-term effective pharmacological treatment modalities are needed, with nil to minimum side-effects. Pharmacological interventions also aid and sustain the effects of a regulated diet and exercise. Moreover, anti-obesity drug research is yet to tailor personalized therapy for better clinical outcomes. Thus, personalized pharmacotherapy, in addition to lifestyle management, seems to be an appropriate multimodality approach in the treatment of obesity and its comorbidities.

Acknowledgments Dr. Gorain would like to acknowledge School of Pharmacy, Taylor's University, Selangor, Malaysia, and Dr. Choudhury, Dr. Pandey and Dr. Verma would like to acknowledge School of Pharmacy, International Medical University, Malaysia for providing resources and support in completing this work.

References

Al-Kuraishy HM, Al-Gareeb AI and Al-Buhadilly AK (2018) Rosuvastatin improves vaspin serum levels in obese patients with acute coronary syndrome. Diseases 6(1). MDPI AG: 9. DOI: 10.3390/diseases6010009

Allison D, Gadde K, Garvey W, Peterson C, Schwiers M, et al. (2012) Controlled-release phentermine/topiramate in severely obese adults: A randomized controlled trial (EQUIP). Obesity 20 (2): 330-342. DOI: doi.wiley.com/10.1038/oby.2011.330

Apovian CM, Aronne LJ, Bessesen DH, et al. (2015) Pharmacological management of obesity: an endocrine society clinical practice guideline. The Journal of Clinical Endocrinology & Metabolism 100(2). Oxford University Press: 342–362. DOI: https://doi.org/10.1210/jc.2014-3415

Aronne LJ, Wadden TA, Peterson C, et al. (2013) Evaluation of phentermine and topiramate versus phentermine/topiramate extended-release in obese adults. Obesity 21(11): 2163–2171. DOI: https://doi.org/10.1002/oby.20584

Astrup A, Rössner S, Van Gaal L, et al. (2009) Effects of liraglutide in the treatment of obesity: a randomised, double-blind, placebo-controlled study. The Lancet 374(9701): 1606–1616. DOI: https://doi.org/10.1016/S0140-6736(09)61375-1

Astrup A, Carraro R, Finer N, et al. (2012) Safety, tolerability and sustained weight loss over 2 years with the once-daily human GLP-1 analog, liraglutide. International Journal of Obesity 36(6): 843–854. DOI: https://doi.org/10.1038/ijo.2011.158

Barrea L, Pugliese G, Muscogiuri G, et al. (2020) New-generation anti-obesity drugs: naltrexone/bupropion and liraglutide An update for endocrinologists and nutritionists. Minerva Endocrinologica 45(2). Edizioni Minerva Medica: 127–137. DOI: https://doi.org/10.23736/S0391-1977.20.03179-X

Boone JE, Gordon-Larsen P, Adair LS, et al. (2007) Screen time and physical activity during adolescence: longitudinal effects on obesity in young adulthood. The International Journal of Behavioral Nutrition and Physical Activity 4. BioMed Central: 26. DOI: https://doi.org/10.1186/1479-5868-4-26

Borgström B (1988) Mode of action of tetrahydrolipstatin: a derivative of the naturally occurring lipase inhibitor lipstatin. Biochimica et Biophysica Acta (BBA)/Lipids and Lipid Metabolism 962(3): 308–316. DOI: https://doi.org/10.1016/0005-2760(88)90260-3.

Cavaliere H, Floriano I and Medeiros-Neto G (2001) Gastrointestinal side effects of orlistat may be prevented by concomitant prescription of natural fibers (psyllium mucilloid). International Journal of Obesity 25(7): 1095–1099. DOI: https://doi.org/10.1038/sj.ijo.0801645

Chan EW, He Y, Chui CSL, et al. (2013) Efficacy and safety of lorcaserin in obese adults: A meta-analysis of 1-year randomized controlled trials (RCTs) and narrative review on short-term RCTs. Obesity Reviews 14(5): 383–392. DOI: https://doi.org/10.1111/obr.12015

Chang S-H, Stoll CRT, Song J, et al. (2014) The effectiveness and risks of bariatric surgery. JAMA Surgery 149(3): 275. DOI: https://doi.org/10.1001/jamasurg.2013.3654

Colman E (2005) Anorectics on trial: A half century of federal regulation of prescription appetite suppressants. Annals of Internal Medicine 143(5): 380–385. DOI: 143/5/380 [pii]

Colman E, Golden J, Roberts M, et al. (2012) The FDA's assessment of two drugs for chronic weight management. New England Journal of Medicine 367(17): 1577–1579. DOI: https://doi.org/10.1056/NEJMp1211277

Connolly HM, Crary JL, McGoon MD, et al. (1997) Valvular heart disease associated with fenfluramine-phentermine. The New England Journal of Medicine 337(9): 581–588. DOI: https://doi.org/10.1056/NEJM199708283370901

Cosentino G, Conrad AO and Uwaifo GI (2013) Phentermine and topiramate for the management of obesity: A review. Drug Design, Development and Therapy 7: 267–278. DOI: https://doi.org/10.2147/DDDT.S31443

Davidson MH, Hauptman J, DiGirolamo M, et al. (1999) Weight control and risk factor reduction in obese subjects treated for 2 years with orlistat: a randomized controlled trial. JAMA 281(3): 235–242. DOI: https://doi.org/10.1001/jama.281.3.235

Davidson MH, Tonstad S, Oparil S, et al. (2013) Changes in cardiovascular risk associated with phentermine and topiramate extended-release in participants with comorbidities and a body mass index $\geq$27 kg/m^2. American Journal of Cardiology 111(8): 1131–1138. DOI: https://doi.org/10.1016/j.amjcard.2012.12.038

Department of Health and Human Services F and, Drug Administration & Center for Drug Evaluation and and Research (CDER). (2007) Guidance for Industry Developing Products for Weight Management. FDA

Després J-P, Golay A, Sjöström L (2005) Effects of rimonabant on metabolic risk factors in overweight patients with dyslipidemia. The New England Journal of Medicine 353(20): 2121–2134. DOI: https://doi.org/10.1056/NEJMoa044537

Dombrowski SU, Knittle K, Avenell A, et al. (2014) Long term maintenance of weight loss with non-surgical interventions in obese adults: systematic review and meta-analyses of randomised controlled trials. BMJ (Clinical research ed.) 348: g2646

Donnelly JE, Blair SN, Jakicic JM, et al. (2009) Appropriate physical activity intervention strategies for weight loss and prevention of weight regain for adults. Medicine & Science in Sports & Exercise 41(2): 459–471. DOI: https://doi.org/10.1249/MSS.0b013e3181949333

Erdmann J, Lippl F, Klose G, et al. (2004) Cholesterol lowering effect of dietary weight loss and orlistat treatment - Efficacy and limitations. Alimentary Pharmacology and Therapeutics 19(11): 1173–1179. DOI: https://doi.org/10.1111/j.1365-2036.2004.01966.x

Fidler MC, Sanchez M, Raether B, et al. (2011) A one-year randomized trial of lorcaserin for weight loss in obese and overweight adults: The BLOSSOM trial. The Journal of Clinical Endocrinology & Metabolism 96(10): 3067–3077. DOI: https://doi.org/10.1210/jc.2011-1256

Finer N, James WP, Kopelman PG, et al. (2000) One-year treatment of obesity: a randomized, double-blind, placebo-controlled, multicentre study of orlistat, a gastrointestinal lipase inhibitor. International Journal of Obesity and Related Metabolic Disorders 24(3): 306–313. DOI: https://doi.org/10.1038/sj.ijo.0801128

Flegal KM, Carroll MD, Kit BK, et al. (2012) Prevalence of obesity and trends in the distribution of body mass index among US adults, 1999-2010. JAMA 307(5): 491. DOI: https://doi.org/10.1001/jama.2012.39

Gadde KM, Allison DB, Ryan DH, et al. (2011) Effects of low-dose, controlled-release, phentermine plus topiramate combination on weight and associated comorbidities in overweight and obese adults (CONQUER): a randomised, placebo-controlled, phase 3 trial. Lancet 377(9774): 1341–1352

Gadde KM, Kopping MF, Wagner HR, et al. (2012) Zonisamide for weight reduction in obese adults. Archives of Internal Medicine 172(20): 1557. DOI: https://doi.org/10.1001/2013.jamainternmed.99

Garvey WT, Mechanick JI (2016) AACE/ACE Guidelines American Association of Clinical Endocrinologists and American College of Endocrinology Clinical Practice Guidelines for Comprehensive Medical Care of Patients with Obesity – Executive Summary. DOI: https://doi.org/10.4158/EP161365.GL

GBD 2015 Obesity Collaborators (2017) Health effects of overweight and obesity in 195 countries over 25 Years. New England Journal of Medicine 377(1): 13–27. DOI: https://doi.org/10.1056/NEJMoa1614362

Golden A (2017) Current pharmacotherapies for obesity: A practical perspective. Journal of the American Association of Nurse Practitioners 29(S1): S43–S52. DOI: https://doi.org/10.1002/2327-6924.12519

Graff SK, Mario FM, Ziegelmann P, et al. (2016) Effects of orlistat vs. metformin on weight loss-related clinical variables in women with PCOS: systematic review and meta-analysis. International Journal of Clinical Practice 70(6). Int J Clin Pract: 450–461. DOI: https://doi.org/10.1111/ijcp.12787

Greco E, Winquist A, Jacob Lee T, et al. (2017) The role of source of protein in regulation of food intake, satiety, body weight and body composition. Journal of Nutritional Health and Food Engineering 6(6): 223. DOI: https://doi.org/10.15406/jnhfe.2017.06.00223

Guyenet SJ, Schwartz MW (2012) Regulation of food intake, energy balance, and body fat mass: implications for the pathogenesis and treatment of obesity. The Journal of Clinical Endocrinology & Metabolism 97(3): 745–755. DOI: https://doi.org/10.1210/jc.2011-2525

Halpern B, Mancini MC (2017) Safety assessment of combination therapies in the treatment of obesity: Focus on naltrexone/bupropion extended release and phentermine-topiramate extended release. Expert Opinion on Drug Safety 16(1): 27–39. DOI: https://doi.org/10.1080/14740338.2017.1247807

Hampp C, Kang EM, Borders-Hemphill V (2013) Use of prescription antiobesity drugs in the United States. Pharmacotherapy: The Journal of Human Pharmacology and Drug Therapy 33 (12): 1299–1307. DOI: https://doi.org/10.1002/phar.1342

Henness S, Perry CM (2006) Orlistat: a review of its use in the management of obesity. Drugs 66 (12): 1625–1656

Hollander P, Bays HE, Rosenstock J, et al. (2017) Coadministration of canagliflozin and phentermine for weight management in overweight and obese individuals without diabetes: A randomized clinical trial. Diabetes Care 2017: 632–639. DOI: https://doi.org/10.2337/dc16-2427

Hong K, Herrmann K, Dybala C, et al. (2016) Naltrexone/Bupropion extended release-induced weight loss is independent of nausea in subjects without diabetes. Clinical Obesity 6: 305–312. DOI: https://doi.org/10.1111/cob.12157

Hu FB, Li TY, Colditz GA, et al. (2003) Television watching and other sedentary behaviors in relation to risk of obesity and type 2 diabetes mellitus in women. JAMA 289(14): 1785. DOI: https://doi.org/10.1001/jama.289.14.1785

Ibrahim N, Moy FM, Awalludin IAN, et al. (2014) The health-related quality of life among pre-diabetics and its association with body mass index and physical activity in a semi-urban community in Malaysia- a cross sectional study. BMC Public Health 14(1): 298. DOI: https://doi.org/10.1186/1471-2458-14-298

James WPT, Caterson ID, Coutinho W, et al. (2010) Effect of Sibutramine on Cardiovascular Outcomes in Overweight and Obese Subjects. New England Journal of Medicine 363(10): 905–917. DOI: https://doi.org/10.1056/NEJMoa1003114

Jensen MD, Ryan DH, Apovian CM, et al. (2014) 2013 AHA/ACC/TOS guideline for the management of overweight and obesity in adults: a report of the American College of Cardiology/American Heart Association Task Force on Practice Guidelines and The Obesity Society. Circulation 129: S102-38. DOI: https://doi.org/10.1161/01.cir.0000437739.71477.ee

Johansson K, Sundström J, Neovius K, et al. (2010) Long-term changes in blood pressure following orlistat and sibutramine treatment: A meta-analysis. Obesity Reviews 11(11): 777–791. DOI: https://doi.org/10.1111/j.1467-789X.2009.00693.x

Jordan J, Astrup A, Engeli S, et al. (2014) Cardiovascular effects of phentermine and topiramate. Journal of Hypertension 32(6): 1178–1188. DOI: https://doi.org/10.1097/HJH.0000000000000145

Kang JG, Park C-Y, Kang JH, et al. (2010) Randomized controlled trial to investigate the effects of a newly developed formulation of phentermine diffuse-controlled release for obesity. Diabetes, Obesity and Metabolism 12(10): 876–882. DOI: https://doi.org/10.1111/j.1463-1326.2010.01242.x

Karmali S, Brar B, Shi X, et al. (2013) Weight recidivism post-bariatric surgery: A systematic review. Obesity Surgery 23(11): 1922–1933. DOI: https://doi.org/10.1007/s11695-013-1070-4

Kashyap SR, Bhatt DL, Wolski K, et al. (2013) Metabolic effects of bariatric surgery in patients with moderate obesity and type 2 diabetes: Analysis of a randomized control trial comparing surgery with intensive medical treatment. Diabetes Care 36(8): 2175–2182. DOI: https://doi.org/10.2337/dc12-1596

Kelly EM, Tungol AA and Wesolowicz LA (2013) Formulary management of 2 new agents: Lorcaserin and phentermine/topiramate for weight loss. Journal of Managed Care Pharmacy 19 (8). Academy of Managed Care Pharmacy (AMCP): 642–654. DOI: https://doi.org/10.18553/jmcp.2013.19.8.642

Kirkham TC (2009) Cannabinoids and appetite: Food craving and food pleasure. International Review of Psychiatry 21: 163–171. DOI: https://doi.org/10.1080/09540260902782810

Lam DD, Garfield AS, Marston OJ, et al. (2010) Brain serotonin system in the coordination of food intake and body weight. Pharmacology Biochemistry and Behavior 97(1): 84–91. DOI: https://doi.org/10.1016/j.pbb.2010.09.003

Lean M, Brosnahan N, McLoone P, et al. (2013) Feasibility and indicative results from a 12-month low-energy liquid diet treatment and maintenance programme for severe obesity. British Journal of General Practice 63(607): 115–124. DOI: https://doi.org/10.3399/bjgp13X663073

Ghee KL (2016) A review of adult obesity research in Malaysia. The Medical Journal of Malaysia 71(Suppl 1): 1–19

Lim YJ, A. R. NA, Minhat HS, et al. (2014) Contributing factors toward social problems among adolescents in governmental rehabilitation center in Kuala Lumpur. International Journal of Public Health and Clinical Sciences 1(1): 70–79

Lonneman DJ, Rey JA and McKee BD (2013) Phentermine/Topiramate extended-release capsules (Qsymia) for weight loss. P and T 38(8): 446. https://www.ncbi.nlm.nih.gov/pmc/articles/PMC3814438/; accessed on 2nd August 2020

Loveman E, Frampton G, Shepherd J, et al. (2011) The clinical effectiveness and cost-effectiveness of long-term weight management schemes for adults: a systematic review. Health Technology Assessment 15(2): 1–182. DOI: https://doi.org/10.3310/hta15020

Luque CA and Rey JA (2002) The discovery and status of sibutramine as an anti-obesity drug. European Journal of Pharmacology 440(2–3): 119–128. DOI: https://doi.org/10.1016/S0014-2999(02)01423-1

Madhu D, Hammad M, Kavalakatt S, et al. (2018) GLP-1 analogue, exendin-4, modulates mapks activity but not the heat shock response in human HepG2 cells. Proteomics - Clinical Applications 12(1): 1600169. DOI: https://doi.org/10.1002/prca.201600169

Malik VS, Popkin BM, Bray GA, et al. (2010) Sugar-sweetened beverages, obesity, type 2 diabetes mellitus, and cardiovascular disease risk. Circulation 121(11): 1356–1364. DOI: https://doi.org/10.1161/CIRCULATIONAHA.109.876185

Manning S, Pucci A, Finer N (2014) Pharmacotherapy for obesity: novel agents and paradigms. Therapeutic Advances in Chronic Disease 5(3): 135–148. DOI: https://doi.org/10.1177/2040622314522848

Marso SP, Daniels GH, Brown-Frandsen K, et al. (2016) Liraglutide and cardiovascular outcomes in type 2 diabetes. New England Journal of Medicine 375(4): 311–322. DOI: https://doi.org/10.1056/NEJMoa1603827

McClendon KS, Riche DM, Uwaifo GI (2009) Orlistat: current status in clinical therapeutics. Expert opinion on drug safety 8(6): 727–744

McNeely W, Goa KL (1998) Sibutramine. A review of its contribution to the management of obesity. Drugs 56(6): 1093–1124. DOI: https://doi.org/10.2165/00003495-199856060-00019

Mehta A, Marso SP, Neeland IJ (2017) Liraglutide for weight management: a critical review of the evidence. Obesity Science & Practice 3(1): 3–14. DOI: https://doi.org/10.1002/osp4.84

Morrato EH, Allison DB (2012) FDA approval of obesity drugs: a difference in risk-benefit perceptions. JAMA 308(11): 1097–1098. DOI: https://doi.org/10.1001/jama.2012.10007

Mozaffarian D, Hao T, Rimm EB, et al. (2011) Changes in diet and lifestyle and long-term weight gain in women and men. New England Journal of Medicine 364(25): 2392–2404. DOI: https://doi.org/10.1056/NEJMoa1014296

Muls E, Kolanowski J, Scheen A (2001) The effects of orlistat on weight and on serum lipids in obese patients with hypercholesterolemia: a randomized, double-blind, placebo-controlled, multicentre study. International Journal of Obesity 25(11): 1713–1721. DOI: https://doi.org/10.1038/sj.ijo.0801814

O'Neil PM, Smith SR, Weissman NJ, et al. (2012) Randomized placebo-controlled clinical trial of lorcaserin for weight loss in type 2 diabetes mellitus: The BLOOM-DM study. Obesity 20(7): 1426–1436. DOI: https://doi.org/10.1038/oby.2012.66

O'Rahilly S, Farooqi IS (2008) Human obesity: a heritable neurobehavioral disorder that is highly sensitive to environmental conditions. Diabetes 57(11): 2905–2910. DOI: https://doi.org/10.2337/db08-0210

Pamuk B, Yilmaz H, Kebapçilar L, et al. (2018) The effect of orlistat and weight loss diet on plasma ghrelin and obestatin. Journal of Research in Medical Sciences 23(1). Wolters Kluwer Medknow Publications. DOI: https://doi.org/10.4103/jrms.JRMS_928_17

Patel SR, Hu FB (2008) Short sleep duration and weight gain: A systematic review. Obesity 16(3): 643–653. DOI: https://doi.org/10.1038/oby.2007.118

Paz-Filho G, Wong M-L, Licinio J (2011) Ten years of leptin replacement therapy. Obesity Reviews 12(5): e315–e323. DOI: https://doi.org/10.1111/j.1467-789X.2010.00840.x

Pi-Synyer FX, Aronne LJ, Heshmati HM, et al. (2006) Effect of rimonabant, a cannabinoid-1 receptor blocker, on weight and cardiometabolic risk factros in overweight or obese patients: RIO-North American: A randomized controlled trial. JAMA 295(7): 761–775

Plieth J (2008) Obesity: what next after the CB1 antagonists' failure. Scrip 2008: 44–47

Plodkowski RA, Nguyen Q, Sundaram U, et al. (2009) Bupropion and naltrexone: A review of their use individually and in combination for the treatment of obesity. Expert Opinion on Pharmacotherapy 10(6): 1069–1081. DOI: https://doi.org/10.1517/14656560902775750

Rao S, Pandey A, Garg S, et al. (2019) Effect of exercise and pharmacological interventions on visceral adiposity: a systematic review and meta-analysis of longterm randomized controlled trials. Mayo Clinic Proceedings 94(2). Elsevier Ltd: 211–224. DOI: https://doi.org/10.1016/j.mayocp.2018.09.019

Reilly JJ, Armstrong J, Dorosty AR, et al. (2005) Early life risk factors for obesity in childhood: cohort study. BMJ (Clinical research ed.) 330(7504): 1357. DOI: https://doi.org/10.1136/bmj.38470.670903.E0

Rodgers RJ, Tschop MH, Wilding JPH (2012) Anti-obesity drugs: past, present and future. Disease Models & Mechanisms 5(5): 621–626. DOI: https://doi.org/10.1242/dmm.009621

Rogge MM, Gautam B (2017) Biology of obesity and weight regain: Implications for clinical practice. Journal of the American Association of Nurse Practitioners 29(S1): S15–S29. DOI: https://doi.org/10.1002/2327-6924.12504

Rössner S, Sjöström L, Noack R, et al. (2000) Weight loss, weight maintenance, and improved cardiovascular risk factors after 2 years treatment with orlistat for obesity. European Orlistat Obesity Study Group. Obesity research 8(1): 49–61. DOI: https://doi.org/10.1038/oby.2000.8

Saunders KH, Shukla AP, Igel LI, et al. (2016) Pharmacotherapy for Obesity. Endocrinology and Metabolism Clinics of North America 45(3): 521–538. DOI: https://doi.org/10.1016/J.ECL.2016.04.005

Scheen AJ, Finer N, Hollander P, et al. (2006) Efficacy and tolerability of rimonabant in overweight or obese patients with type 2 diabetes: a randomised controlled study. Lancet 368(9548): 1660–1672. DOI: https://doi.org/10.1016/S0140-6736(06)69571-8

Secher A, Jelsing J, Baquero AF, et al. (2014) The arcuate nucleus mediates GLP-1 receptor agonist liraglutide-dependent weight loss. Journal of Clinical Investigation 124(10): 4473–4488. DOI: https://doi.org/10.1172/JCI75276

Shapiro M, Reid A, Olsen B, et al. (2016) Topiramate, zonisamide and weight loss in children and adolescents prescribed psychiatric medications: A medical record review. International Journal of Psychiatry in Medicine 51(1). SAGE Publications Inc.: 56–68. DOI: https://doi.org/10.1177/0091217415621266

Sherman MM, Ungureanu S, Rey JA (2016) Naltrexone/Bupropion ER (Contrave): Newly approved treatment option for chronic weight management in obese adults. P&T 41(3): 164–172. https://pubmed.ncbi.nlm.nih.gov/26957883/; accessed on 2nd August 2020

Shin JH, Gadde KM (2013) Clinical utility of phentermine/topiramate (QsymiaTM) combination for the treatment of obesity. Diabetes, Metabolic Syndrome and Obesity: Targets and Therapy 6: 131–139. DOI: https://doi.org/10.2147/DMSO.S43403

Sjostrom L, Rissanen A, Andersen T, et al. (1998) Randomised placebo-controlled trial of orlistat for weight loss and prevention of weight regain in obese patients. European Multicentre Orlistat Study Group. Lancet 352(9123): 167–172. DOI: S0140673697115094 [pii]

Sjöström L, Narbro K, Sjöström CD, et al. (2007) Effects of bariatric surgery on mortality in Swedish Obese subjects. New England Journal of Medicine 357(8): 741–752. DOI: https://doi.org/10.1056/NEJMoa066254

Sjöström L, Peltonen M, Jacobson P, et al. (2012) Bariatric surgery and long-term cardiovascular events. JAMA 307(1): 56. DOI: https://doi.org/10.1001/jama.2011.1914

Smith SR, Weissman NJ, Anderson CM, et al. (2010) Multicenter, placebo-controlled trial of lorcaserin for weight management. New England Journal of Medicine 363(3): 245–256. DOI: https://doi.org/10.1056/NEJMoa0909809

Smith SR, Garvey WT, Greenway FL, et al. (2015) Combination weight management pharmacotherapy with lorcaserin and immediate release phentermine. Journal of General Internal Medicine 30: S124

Speakman JR, Levitsky DA, Allison DB, et al. (2011) Set points, settling points and some alternative models: theoretical options to understand how genes and environments combine to regulate body adiposity. Disease Models & Mechanisms 4(6): 733–745. DOI: https://doi.org/10.1242/dmm.008698

Srivastava G, Apovian CM (2018) Current pharmacotherapy for obesity. Nature Reviews Endocrinology 14(1): 12–24. DOI: https://doi.org/10.1038/nrendo.2017.122

Stanford FC (2019) Controversial issues: A practical guide to the use of weight loss medications after bariatric surgery for weight regain or inadequate weight loss. Surgery for Obesity and Related Diseases 15(1). Elsevier Inc.: 128–132. DOI: https://doi.org/10.1016/j.soard.2018.10.020

Swinburn BA, Sacks G, Hall KD, et al. (2011) The global obesity pandemic: shaped by global drivers and local environments. Lancet (London, England) 378(9793): 804–814. DOI: https://doi.org/10.1016/S0140-6736(11)60813-1

Taylor JR, Dietrich E, Powell J (2013) Lorcaserin for weight management. Diabetes, Metabolic Syndrome and Obesity 6: 209–216. DOI: https://doi.org/10.2147/DMSO.S36276

Torgerson JS, Hauptman J, Boldrin MN, et al. (2004) XENical in the prevention of diabetes in obese subjects (XENDOS) study: a randomized study of orlistat as an adjunct to lifestyle changes for the prevention of type 2 diabetes in obese patients. Diabetes Care 27(1): 155–161

Van Gaal LF, Rissanen AM, Scheen AJ, et al. (2005) Effects of the cannabinoid-1 receptor blocker rimonabant on weight reduction and cardiovascular risk factors in overweight patients: 1-Year experience from the RIO-Europe study. Lancet 365(9468): 1389–1397. DOI: https://doi.org/10.1016/S0140-6736(05)66374-X

Wadden TA, Webb VL, Moran CH, et al. (2012) Lifestyle modification for obesity: New developments in diet, physical activity, and behavior therapy. Circulation 125(9): 1157–1170. DOI: https://doi.org/10.1161/CIRCULATIONAHA.111.039453

Wang GJ, Tomasi D, Volkow ND, et al. (2014) Effect of combined naltrexone and bupropion therapy on the brain's reactivity to food cues. International Journal of Obesity 38(5): 682–688. DOI: https://doi.org/10.1038/ijo.2013.145

Weintraub M, Sundaresan PR, Madan M, et al. (1992) Long-term weight control study I (weeks 0 to 34): The enhancement of behavior modification, caloric restriction, and exercise by fenfluramine plus phentermine versus placebo. Clinical Pharmacology & Therapeutics 51(5): 586–594. DOI: https://doi.org/10.1038/clpt.1992.69

Weissman NJ, Smith SR, Fain R, et al. (2017) Effects of lorcaserin on pre-existing valvulopathy: A pooled analysis of phase 3 trials. Obesity 25(1): 39–44. DOI: https://doi.org/10.1002/oby.21695

Wharton S (2016) Current perspectives on long-term obesity pharmacotherapy. Canadian Journal of Diabetes 40(2): 184–191. DOI: https://doi.org/10.1016/j.jcjd.2015.07.005

Whitaker RC (2004) Predicting pre-schooler obesity at birth: the role of maternal obesity in early pregnancy. Pediatrics 114(1): e29-36

Wilding J (2007) Clinical investigations of antiobesity drugs. In: Appetite and Body Weight 337–355. DOI: https://doi.org/10.1016/B978-012370633-1/50014-1

Yanovski SZ, Yanovski JA (2014a) Long-term drug treatment for obesity: A systematic and clinical review. JAMA 311(1): 74–86. DOI: https://doi.org/10.1001/jama.2013.281361

Yanovski SZ, Yanovski JA (2014b) Naltrexone extended release plus bupropion extended release for treatment of obesity. JAMA 313(12): 1213–1214. DOI: https://doi.org/10.1001/jama.2015.1617

Yumuk V, Tsigos C, Fried M, et al. (2015) European guidelines for obesity management in adults. Obesity Facts 8(6): 402–424. DOI: https://doi.org/10.1159/000442721

Zhou Y-H, Ma X-Q, Wu C, et al. (2012) Effect of anti-obesity drug on cardiovascular risk factors: a systematic review and meta-analysis of randomized controlled trials. PLoS ONE 7(6): e39062. DOI: https://doi.org/10.1371/journal.pone.0039062

Nanotechnology in the Context of Obesity 11

Satheesh Babu Natarajan, Suriyakala Perumal Chandran, and Methil Kannan Kutty

Abstract

Obesity is a pandemic disease with immense magnitude and it is due to escalating death with co-morbidity in this century. Obesity is caused by combinatorial factors including genetic, metabolic, behavioral, environmental, cultural, and socioeconomic factors that contribute to a person's body weight. This condition parse is alarmingly big magnitude problem and it is conformingly with the co-morbidities. An elevated body mass index (BMI) increases the prevalence, morbidity, and mortality of type 2 diabetes mellitus, hypertension, heart disease, stroke, osteoarthritis, respiratory tract disorders, gallstones, certain types of cancer, and psychological disorders. It is striking that only a few therapeutic agents are available to treat obesity including orlistat, sibutramine, lorcaserin, phentermine-topiramate available to treat obesity. These agents can reduce body weight by decreasing the consumption or absorption of food or by augmenting energy disbursement. In recent decades, the limelight of nanotechnology has opened numerous new vistas in medical sciences, especially in the area of drug delivery. The nanoparticulate drug delivery system can transport specific anti-obesity drugs to the white adipose tissue in the body, aiding to evade potential side effects that can occur if the drugs find their way to other parts of the body. In this chapter, we highlighted the various nano-drug delivery systems including

S. B. Natarajan (✉)
Department of Pharmaceutics, Faculty of Pharmacy, Lincoln University College, Petaling Jaya, Malaysia

S. Perumal Chandran
Department of Biochemistry, Faculty of Medicine, Lincoln University College, Petaling Jaya, Malaysia

M. K. Kutty
Department of Pathology, Faculty of Medicine, Lincoln University College, Petaling Jaya, Selangor, Malaysia

M. K. Kutty, A. Elengoe (eds.), *Obesity and its Impact on Health*,
https://doi.org/10.1007/978-981-33-6408-0_11

polymeric nanoparticles, chitosan nanoparticles, Polyethylene glycol-based nanoparticles, gold nanoparticles, liposomes, transfersomes, and microparticles to enhance the therapeutic efficacy against obesity treatment.

Keywords

Obesity · Nanotechnology · Drug delivery system

11.1 Introduction

Obesity is a pandemic disease with immense magnitude and it's due to escalating death with co-morbidity. Obesity is defined as substantive adiposities with co-morbidities including hypertension, diabetes, etc. As a result of an imbalance between lipolysis and lipogenesis; that leads to excessive accumulation of fats in adipocytes (Jyotibala 2017; Chandrasekaran et al. 2012). Around 60% of adult peoples are suffering from obesity with high body mass index (BMI) (National Center for Health Statistics 1999). No one, irrespective of age, sex, race, and ethnicity is immune to this salvage (US Department of Health and Human Services 2004). Many factors strongly influence the cause of obesity including genetic, behavioral, metabolic, cultural, environmental, cultural, and socioeconomic factors. This condition parse is alarmingly big magnitude problem and it is conformingly with the co-morbidities. An augmented BMI raised the prevalence, mortality and morbidity of chronic diseases (Overweight, Obesity, and Health Risk 2000). There is proof to demonstrate that histamine, serotonin, dopamine, and their associated receptor activities regulate obesity (Chandrasekaran et al. 2012). However, the development of obesity, significant shreds of evidence needs to establish the function of ghrelin, leptin, and neuropeptides (Berbari et al. 2013; Khazaei and Tahergorabi 2013; Singer et al. 2013).

11.2 Treatment Strategy

It is striking that only a few therapeutic agents are available to treat obesity including orlistat, sibutramine, lorcaserin, phentermine-topiramate available to treat obesity. These agents may control the body weight by controlled eating habits or promoting maximizes energy expenditure (Cooke and Bloom 2006; Sargent and Moore 2009). Regrettably, the above-mentioned medicines have not proven a prominent success rate to control body weight and however, the majority of therapeutic agents are not accessible for obesity therapy owing to their rigorous adverse effect (Padwal and Majumdar 2007). A pancreatic and gastric lipase inhibitor (e.g. Orlistat) utilized to treat obesity as a long-term therapy; and selective serotonin 2C receptor against Lorcaserin that principally controls the consumption of food.

While heart and heart valves are expressed for serotonin 2C receptors agonist existence of receptors for fenfluramine and dexfenfluramine drugs are lacking. Recently FDA approved the combination of phentermine and topiramate (PHEN/

TPM) to induce central norepinephrine release and promote weight reduction by decreasing food ingestion respectively. Adverse side effects have led to disapproved by the FDA (fenfluramine, dexfenfluramine, phenylpropylamine) or withdrawn voluntarily (sibutramine). In US clinical research group has invented many new anti-obesity drugs (taranabant, rimonabant, high-dose topiramate) to combat obesity but these agents are not approved due to their potential side-effects.

11.3 Nanotechnology in Medicine

Nanoparticle is defined as any material with a dimensional range of approximately 1–100 nanometer (nm) and exhibits chemical, physical, and biological effects properties (Dowling and Grobert 2011; FDA 2012). In the last 3–4 decades observed that tremendous advancement in the field of nanotechnology especially in the medical and pharmaceutical industry. Numerous synthetic and biological products are entering into the medical industry to treat many acute and chronic diseases. However, the majority of these compounds faced difficulties to reach biological systems to achieve expected pharmacological action due to their incompatibilities and complex physio-chemical properties. The emergence of the current nanotechnological approach permits to progress significantly to attain sustainable and controlled drug delivery at the desired site. The exponential market volume of nanotechnology-based drug delivery systems will be benefited by the pharmaceutical companies and market players. The augmented numbers of patented products have evidenced the acceptance and usage of nanotechnology in the field of medicine. Numerous terminologies have utilised to define and function of nano-drug delivery systems. In majority, either lipids or biodegradable polymers are commonly used as carrier material to encapsulate or entrap the drug molecule(s), and the expected tie achieves the particle size of nanoparticles from few nanometers to a few hundred nanometers range.

11.4 Rationale for Nanotechnology for Obesity

The recent advancement to treat obesity includes strict control of dietary intake; excellent energy dispersement by increased physical exercise; pharmaceutical approach such as Orlistat, Lorcaserin, and Phentermine-topiramate; and surgical technique such as gastric by-pass surgical method. In recent times the only little number of drugs is existed to treat obesity due to the complex interplay of genetic, cultural and environmental factors (Friedman 2009; Colman 2012; Barnhart 2011). Those drugs are used to reducing calorie consumption by either reducing nutritional metabolism or suppressing hunger by stimulating the central nervous system. Conversely, these drugs only grant humble pharmacological action and are regrettably produced objectionable and significantly deleterious side-effects. Therefore, unconventional approaches with the extensive investigation are needed to be executed to fight against obesity (Chunhui et al. 2015).

A non-invasive mode of delivery is also recommended for combating obesity. Capsaicin is one of the remarkable drugs that have been exhibited to increase the metabolic rate and decrease obesity in mouse models. Though capsaicin needs a relatively higher dose; but sustained release behavior to attain the anti-obesity impacts. Moreover, capsaicin is not recognized to localize to a specific targeted site or specific cells, therefore dosing needs to increase to attain the anti-obesity effect. Also, the long term administration of capsaicin has produced diverse side-effects, which may impact the poor patient-compliance along with GIT discomfort and esophageal irritation (US 20150190531 A1).

A novel targeted delivery approach has been utilized to achieve drug release at adipose tissue and is considered as a successful modality to treat metabolic disorders (Hossen et al. 2012). Numerous prospective targeted sites of adipose tissue have been identified and anti-obesity agents have invented unfortunately many of these agents failed to reach the targeted site of adipose tissue, consequently achieve poor efficacy and produce unwanted side-effects.

Conversely, protein-based drugs are well recognized for the treatment of obesity however, the systemic circulation of proteins failed to achieve optimum serum bioavailability due to their inefficient to pass through the blood-brain barrier (BBB). Intensive investigations have been executed to enhance the protein concentration in the brain region to treat numerous diseases of the central nervous system (CNS). Intranasal delivery has currently been explored as an efficient approach to deliver proteins and therapeutic agents to the brain and thus, this approach permits by-passing the BBB and minimize the peripheral tissue exposure (Boer and Gaillard 2007; Lichota et al. 2010; Gabathuler 2010).

In addition, there is a mounting interest in the utilization of nanoparticles in the CNS delivery to enhance the stability and pharmacokinetic profile of proteins by encapsulating them into nanoparticulates and further augmented their brain uptake by BBB-specific targeting moieties (Tiwari and Amiji 2006; Tosi et al. 2007, 2013). The field of nanotechnology keeps progressing forward to achieve greater considerate of the BBB and obtain skill in formulating various drug delivery systems to treat not only brain diseases and also obesity.

Leptin is a hormone (protein 17-kDa) that is secreted by fat cells that control body adiposity by crossing the BBB to influence feeding and thermogenesis. Particularly, obesity in humans and rodent models is believed to be developed by inefficient transport of leptin across BBB (US Department of Health and Human Services 2004), rendering leptin-targeted drugs are inefficient to the treatment of obesity patients (Van Heek et al. 1997; Banks et al. 1999). Glucagon-like peptide-1 (GLP-1) hormone also poses a significant function to maintain glucose homeostasis. The GLP-1 receptors are found in the gastrointestinal tract, pancreas, heart and brain (Holst 2007). Exenatide and liraglutide (GLP-1 receptor agonistsas) have been recognized for the treatment of type II diabetes and also used to treat obesity with non-diabetic patients (Araújo et al. 2016). Enteral administration of proteins is still facing many hurdles to reach the desirable pharmaceutical and pharmacological efficiency. While nanoparticulate drug delivery system is promising colloidal carrier systems has been proposed to increase bioavailability of protein, however *in-vitro* and *in-vivo* investigation needs to appraise the effectiveness of the recently developed oral protein-loaded nanoparticulate delivery systems.

11.5 Nano Drug Delivery System for the Treatment of Obesity

In the battle against obesity, scientists from the Massachusetts Institute of Technology (MIT) and Brigham and Women's Hospital have developed nanoparticulates that deliver drugs straight to fatty tissue, and escape from other organs and tissues where the drug induce side-effects. This drug delivery system transport the anti-obesity drugs directly to the white adipose tissue, facilitate to evade the possible side-effects which may produce if the therapeutic agents reach the undesired region of the body. This delivery system has been potentially utilized for cancer therapy, particularly targeting the cancerous cells and delivering drugs directly to the desired site. Brown adipose tissue is remarkable body fat that burns away energy for the thermogenesis process. White adipose tissue converts into brown adipose tissue by some drugs; however, there have been hurdles for the drugs to target specifically white fatty tissue. Numerous vascular present in brown adipose tissue as compared with white adipose tissue, hence during the transformation, the nanoparticulate drug delivery system becomes highly reinforced and effective.

11.5.1 Polymeric Nanoparticles

Polymer-based nanoparticulate systems are predominantly formulated by using a biodegradable class of polymers and the highly analyzed to achieve controlled release (Madureira et al. 2015). Polymeric nanoparticles possess to enhance oral bioavailability, achieve improved encapsulation, ability to produce controlled drug release, and minimal toxicity.

Numerous investigation was executed to developed nanoparticles loaded antimicrobial agents using poly lactic- co-glycolic acid (PLGA), polylactic acid (PLA), gelatin, chitosan, poly-alkyl-cyanoacrylates and polycaprolactone to treat gastrointestinal infections (Kumari et al. 2010). Hossen et al. (2012) formulated and investigated the nanoparticles (NP) with short PEG (2 kDa)-spacer loaded with prohibitin-homing peptide ligand to deliver the drug into the cytoplasmic endothelial cells in murine adipose tissue. Diet-induced obesity models were used to investigate the dual-targeting approach with augmented targeting efficiency of nanoparticles to adipose endothelial cells. The targeted accumulation of prohibitin-targeted nanoparticles (PT-NP), modified with a peptide ligand via a long PEG-linker, was significantly increased in white fat vessels of normal healthy mice as compared to non-PEGylated NPs (Hossen et al. 2012). This investigation exhibited that the targeted delivery of PEGylation of nanoparticles is a predominant feature in enhancing the *in-vivo* targeted delivery of nanoparticles.

This investigation revealed that the systemic delivery of PT-NP targeted to both adipose tissues and angiogenic clusters of obese fatty cells into DIO mice.

Remarkably without targeting moieties, PEGylated NPs also accumulated in these clusters, which was due to the nanosized carriers passively accumulate in clusters through EPR (enhanced permeability and retention) mechanism. This research finding demonstrated that this delivery system potentially useful for the development of vascular targeted nanomedicine to treat obesity (Hossen et al. 2012).

11.5.2 Chitosan Nanoparticles

Chitosan is a type oligosaccharide (COS) of β-1, 4 linked D- glucosamine generally prepared by deacetylation and hydrolysis of chitin. It is highly aqueous soluble, non-carcinogenic, easily absorbed through the intestinal region and excreted through the kidney. The current investigation explained that COS has possessed many pharmacological activities including antioxidant, anti-inflammatory, neuroprotective, anti-cancer, antibacterial, hypoglycemic and liver protective activity. Additionally, COS is also well-utilized in the food industry to control or minimize lipidic levels and act as an efficient lipid-lowering nutritional ingredient.

Murine Swiss cells from embryos (3 T3-L1 cell line) models are well documented and used to investigate the efficiency of low molecular weight chitosan (1–3 kDa) on inhibition of adipocyte differentiation (Kumar et al. 2009; Rahman et al. 2008).

Chitosan remarkably reduced the expression of adipogenic transcription factors (such as peroxisome proliferator-activated receptor γ (PPARγ)) and also considerably down-regulated adipogenic marker proteins (such as leptin, adiponectin and resistin) (Cho et al. 2008). However, the anti-obesity mechanism of chitosan has not been extensively investigated. In this research, the nutrition-induced animal model has been used to demonstrate the anti-obesity efficiency of glucosamine (GA), COS-1, (average molecular weight $\leq$ 1000 and COS-2 (average molecular weight $\leq$ 3000). The authors analyzed the effect of chitosan molecular weight and glucosamine on its anti-obese activity. Potential alterations of gene sequencing treated with GA and COS-1 were analyzed using the digital gene expression tag profiling (DGE) method. Gene ontology and pathway significant functional enrichment studies were executed to identify the potential pathways activated by GA and COS-1 during weight reduction. Moreover, the LXR-α gene (nuclear receptors) is a known direct target of PPARγ which coordinates to regulate lipid metabolism (Chawla et al. 2001; Lanlan et al. 2015).

Tea is one of the most consumed beverages next to the water, which is obtained from *Camellia sinensis* plant widely grown in many countries. Catechin is a flavonols class of phytoconstituent which is abundantly present in tea leaves. Numerous studies evidenced that tea catechins have many health benefits including fighting cancers, diabetes, obesity and cardiovascular diseases. Epigallocatechin Gallate (EGCG) is a class of catechin evidenced with antioxidant, anticancer, anti-obesity activity. In obesity treatment, EGCG primarily promoting lipid metabolism, burn enormous calories and leads to fat loss. However, the poor stability and intestinal absorbance constitute the major setback to utilize this phytoconstituent in the medicinal field. It hypothesized that encapsulation of catechins into nanoparticulates leads to improved stability and maximize therapeutic efficacy. To overcome the intestinal instability, EGCG has been encapsulated into chitosan-tripolyphosphate nanoparticles (E-CNP) and investigate the efficiency of oral absorption of EGCG on mice models. This study revealed that E-CNP improved the plasma concentration of EGCG by nearly two fold as compared with EGCG solution, which confirmed that chitosan-based nanoparticles could be the favored approach to enhance the oral bioavailability of EGCG in many treatment modalities.

11.5.3 PEG-Based Nanoparticles

Polyethylene glycol (PEG) is a linear synthetic polyether the most versatile and famous "stealth" polymer in the drug delivery sector, due to its long history of safety in humans and classification as Generally Regarded as Safe (GRAS) by the FDA. More importantly, it is highly aqueous soluble and biocompatible with numerous therapeutic agents. Surface functionalization or coating of nanoparticles with polyethylene glycol (PEG) is commonly known as "PEGylation". This approach is widely utilized to enhance the therapeutic efficiency of bioactive ingredients and gene delivery to the target site by protecting the surface of the nanoparticles from aggregation, phagocytosis and opsonization process. Davis and Abuchowski were described firstly in 1977 that the covalent attachment of PEG to bovine serum albumin and liver catalase proteins. They revealed that PEGylation of liver catalase shown enhanced systemic circulation duration and lower the immunogenicity of the proteins without compromising pharmacological property.

Hossen et al. (2012) explored the potential of PEGylated-based nanoparticulates to target obese fat through passive targeting mechanisms in cancer therapy (Hossen et al. 2012). The constructive effect of prohibition targeted nanoparticle (PTNP) encapsulated in a proapoptotic peptide (D (KLAKLAK) 2, KLA) on DIO and dysfunctional adipose tissue, a major mediator of the metabolic syndrome, as evidenced by ectopic fat deposition. The systemic injection of diet-induced obese mice with a low dose of KLA-PTNP, rather than a bioconjugate composed of the same targeting peptide and KLA (Adipotide) revealed in an exhibited body weight, as evidenced by a significantly low level of serum leptin, equivalently with an anti-obesity effect on dysfunctional adipose cells, including adipocytes and macrophages. In addition, the KLA-PTNP treatment resulted in a reduction in ectopic fat deposits in the liver and muscle with the lipolytic action of elevated serum adiponectin, with no detectable hepato-toxicity. Notably, PTNP accumulated in obese fat through the EPR effect was enhanced by multivalent active targeting and cytoplasmic delivery into adipose endothelial cells via escaping from endosomes/lysosomes. Thus, the vascular-targeted nano delivery system potentially contributes to the control of adipose function and ectopic fat deposition associated with obesity and the metabolic syndrome (Hossen et al. 2012).

11.5.4 Gold Nanoparticles

Gold nanoparticles (GNP) functionalized with different therapeutic agents have been effectively utilized as controlled drug delivery to treat many diseases, disorders, diagnostic tools in the pharmaceutical industry. Thovhogi et al. (2015) investigated the binding and targeting efficiency of the adipose homing peptide (AHP) functionalized GNP (AHP-GNP) using flow cytometric analysis method and inductively coupled plasma optical emission spectroscopy on $CaCO_2$ cell lines and rats fed either chow (or) high-fat diet and compared the against unfunctionalized GNP. This investigation demonstrated that functionalized AHP-GNP shown greater cellular uptake efficiency than unfunctionalized gold nanoparticles and the binding efficiency was functionalized AHP-GP was followed concentration and temperature-dependent manner.

Targeting potential of functionalized nanoparticles (AHP-GNP) and unfunctionalized nanoparticles (GNP) were tested on obese rats to white adipose tissue. As a result, Thovhogi et al. (2015) concluded with the functionalized AHP-GPNP has a high potential for targeting white adipose tissue in obese rats. From this report, it revealed that GNP has the potential to treat obesity with a targeting delivery mechanism (Thovhogi et al. 2015).

11.5.5 Liposomes

Liposomes are one the prominent vesicular based drug delivery system developed for medicinal applications in many segments, including drug delivery, gene therapy vehicles, vaccine adjuvants, diagnostic kits, and artificial blood (Margalit and Yerushalmi 2005). The liposomes are prepared by suspending phospholipids in an organic solvent and dehydrating the mixture leaving a thin phospholipid film on the bottom of the vessel. Then aqueous phase is added to the mixture and allowed to swell and dispersed by a mechanical stirrer. Liposomes were obtained from the process yield with multilamellar vesicles (MLVs).

Therapeutic aptamers are a class of short single-stranded oligonucleotide ligand that can attach with a broad range of targets with greater binding and specificity and possess great importance in therapeutic, imaging and diagnosis fields (Tuerk and Gold 1990).

Hossen et al. (2012) have formulated adipose vasculature-targeted nanocarrier referred to as prohibitin-targeted-nanoparticle (PTNP) using the peptide ligand. This delivery system can target adipose tissue and efficiently deliver drugs into the adipose endothelial cells. They have also investigated the effective treatment strategy for obesity in mice model using proapoptotic peptide/protein-loaded PTNP.

Liposomes are versatile vesicular delivery system which is commonly used vehicles functionalized with aptamers for targeted drug delivery. The efficiency of a targeted-liposomes surface to target the site was principally based on the free PEG. By utilizing the affinity DNA aptamers, they have formulated the targeted nanoparticle very successfully.

Macrophages play a crucial role in the progression of obesity-related inflammation. Novel therapeutic approach for the delivery of inflammatory compounds such as interleukin 10 (IL 10) to macrophage can enhance the pharmacological profile of many drugs. Toita et al. (2016) also have used phosphatidylserine (PS) as a carrier material to formulate a liposomal delivery system (PS-L), which have potential anti-inflammatory and macrophage targeting capability and also act as a carrier vehicle for the delivery of IL 10 to macrophages. The IL 10-conjugated PS-L (PS-L-IL 10) exhibited higher rate specificity to macrophages. In this investigation, PS-L-IL 10 delivery system exhibited considerable anti-obesity and anti-inflammatory efficacy in obese mice whereas animals treated alone with IL-10 or PSL did not show significant results. These findings recommended that the PS-L-IL 10 has potential targeting efficiency to macrophage and shown augmented anti-inflammatory action (Toita et al. 2016).

Bin Feng et al. (2011) investigated the liposomes loaded clodronate to enhance the metabolic parameter and reduce visceral adipose macrophage content in diet-induced obese mice models. Intra-peritoneal administration of liposomes loaded clodronate prepared to diminish macrophages in lean and diet-induced obese mice models and investigated the macrophage content in visceral adipose tissue, metabolic parameters, glucose and insulin tolerance, histological parameters of liver and adipose, secretion of adipokine and cytokine. This investigation demonstrated that the intraperitoneal administration of liposomal clodronate can decrease visceral adipose tissue macrophages, enhance glucose-homeostasis and insulin sensitivity in diet-induced obese mice models.

Balada et al. (1997) investigated the efficacy body weight of liposomes encapsulated with the oleoyl-estrone (slimming agent) on Zucker obese rats. After 28th day period, the glucose level was maintained; whereas, the triacylglycerides level was improved and total cholesterol was decreased (Balada et al. 1997).

5β-steroids are well known to be very effective in controlling obesity and linked diabetic and/hyper cortical syndrome, these agents administered orally treat these conditions, the rate of absorption of the 5β-steroids is approximately 15% as analyzed by various assays procedure. It revealed that a high percentage of the orally administered drug is never reached the systemic circulation and most of the administered drug is excreted through fecal matter. Joseph and Elliot (1997) developed a liposomal formulation encapsulated with 5β-steroids with (o)r without the OB protein. It is expected that liposome formulation including this 5β steroid and OB protein administered parenterally to utilize the different mechanisms of the two active ingredients to enhance the therapeutic efficacy (Joseph and Elliot 1997).

11.5.6 Cerium Dioxide Nanoparticles

Recently, cerium dioxide-based nanoparticles ($nCeO_2$) have been reported for its anti-oxidant and catalytic activity, which is due to the presence of two valence states (Ce^{3+} and Ce^{4+}). As a result oxygen vacancies exist in these nanoparticulates can react with the neighboring reactive oxygen species (ROS), hence exploring $nCeO_2$

as a prospective *in vivo* mimetic for endogenous anti-oxidants like superoxide dismutase; however, there are still few conflicting statements about the efficiency of $nCeO_2$ anti-oxidant activity. Therefore, antioxidants are very crucial for the successful treatment of liver injury (Dowman et al. 2010).

Another research team Nazarii et al. (2016) also investigated the efficacy of $nCeO_2$ on antioxidant enzyme activity and lipid peroxidation in experimentally induced NAFLD rats. The administration of $nCeO_2$ reduces its pathological effects due to its capability of Ce^{3+} conversion into Ce^{4+} $nCeO_2$ successfully defuse hydrogen peroxide and hydroxyl radical without producing ROS. Lipid peroxidation decreases the pathological effect of ROS in the liver tissue, which is evidenced by the considerable improvement of the main NAFLD histological features and the re-establishment of antioxidant potential. In conclusion, $nCeO_2$ has the potential to be utilized in the drug delivery system for the treatment of NAFLD; however, it should be investigated using other animal models to confirm the anti-obesity mechanism (Nazarii et al. 2016).

11.5.7 Chitosan Microsphere

Capsaicin (CP) is an active compound isolated from pepper which belongs to *Capsicum annuum*. It has a strong characteristic of hot and pungency taste. Reinbach et al. (2009) demonstrated that capsaicin has the potential to enhance satiety, promote energy expenditure and have a tendency to inhibit overeating. Capsaicin loaded chitosan microspheres (C-CM) formulated and analyzed its anti-obesity potential of C-CM in obese induced rat models. Compared with chitosan microspheres, CP, and Orlistat, the C-CM exposed greater potential to control body weight, body mass index, body fat, organ index, the proportion of fat to body weight, and serum lipids. The C-CM upregulated the expressions of UCP2, PPARα, PPARγ, and adiponectin has down-regulated the expression of leptin. Hence, CCMS may be deemed novel, effective, safe and natural weight loss compounds, and there is also an additive effect between capsaicin and chitosan (Tan et al. 2014).

11.5.8 Floating Microspheres

Orlistat is an effective lipase inhibitor and it is used for the treatment of obesity by inhibiting the absorption of dietary fats. Sunil et al. (2006) prepared orlistat (anti-obesity agent) loaded floating microspheres and investigate the gastro-retentive and controlled-release mechanism. Various formulation and process variables were tested on the morphological character, entrapment efficiency, floating efficiency, and drug release behavior was studied. The optimized microsphere formulation was administered orally to rabbits and blood samples were collected to analyze the pharmacokinetic behavior. The optimized orlistate microsphere exhibited good floating capability in simulated gastric fluid and release the drug behavior of orlistat floating microspheres followed Peppas-Korsmeyer and Higuchi matrix model (Sunil et al. 2006).

11.5.9 Sustained Release Microspheres

Orlistat was reported with poor aqueous solubility but freely soluble in ethanol and methanol. Its mode of action is inhibiting gastric and pancreatic lipases; blocking the action of lipase and thereby thwart the breakup and absorption of 25% of the fat in a meal. Fathima et al. (2012) developed microspheres of Orlistat with suitable doses for sustaining the release, which focuses to minimize the dosing frequency, side effects and enhanced bioavailability. The Orlistat microspheres were formulated by the solvent evaporation method with hydrophobic ethylcellulose polymer for retarding the drug release at different drug and polymer ratios in the presence of different surfactants such as tween 80 and pluronic F68. The formulated microspheres were evaluated for its particle size distribution (2–5 μm), drug content (95–99%) and drug entrapment efficiency (99%). The infrared spectroscopy analysis fortified the above data and ratified that drug and polymer were more compatible with each other. The *in-vitro* drug release behavior of the optimized formulation exhibited that it follows Higuchi, Peppas, Hixon, zero-order and first-order kinetics. The Orlistat microspheres are found to have sustained release of drugs exceeded 12 h (Fathima et al. 2012).

Rimonabant is an important class of anti-obesity agent which can reduce the appetite by acting as Cannabinoid-1 receptor antagonist. Sodium alginate and guar gum polymer mixture have been used to prepare drug-loaded microspheres using the ionotropic gelation technique to achieve dose-related side-effects. Gupta et al. (2010) has used guar gum s polymer material to formulate Rimonabant loaded microspheres for gastrointestinal drug delivery for effective obesity therapy. An extensive research investigation is required to predict the specificity of drug uptake in the gastrointestinal site. Sodium alginate is one of the important polymeric materials which permits the microspheres degraded in very slow manner thus achieve slow drug release from the microspheres. Rimonabant was effectively encapsulated in microsphere by optimizing the proportion of sodium alginate and guar gum to encapsulate many pharmaceutical agents to minimize the toxicity and side-effects (Gupta et al. 2010).

11.5.10 Transfersomes

Transfersome is another vesicular drug delivery system explored in recent decades to deliver many therapeutic agents through the transdermal route to achieve augmented penetrate rate. The structural features of transfersome are similar to the biological-membrane and it can perform as a multitalented drug carrier (Shikha et al. 2014). Deformable transfersome also have the capability to penetrate efficiently and pass through various intact dermal layers. Initially in the 1960s, transfersome has developed as a carrier system and in the later 1990s, it has been exploring for clinical application.

Transfersomes have possessed numerous advantages than compared with liposomes including better therapeutic index, used as carriers for numerous pharmacological agents including proteins and peptides and enhancing the amount of drug permeation through the skin. Emodin is a Chinese herbal medicine evidenced to have many pharmacological activities on liver fibrosis, hepatocyte injury and inflammation, etc. Recent research evidence that emodin has the potential to initiate the weight loss of high-fat feed rats, control plasma lipids and blood glucose level; and promote anti-lipid peroxidation. However, emodin has reported poor penetration through the skin which affects its therapeutic efficacy. Hence, there is a need to develop a suitable transdermal drug delivery system for emodin to enhance the penetration potential. The emodin encapsulated transferosomal delivery vehicles have developed by Kun et al. (2014); investigate the anti-obesity profile in adipose tissue of fatty food-induced obese rats. The results depicted that nano emodin transfersome could decrease adipocytes size and body weight gain and insulin sensitivity and transfersomes could reduce the dosage due to its augmented skin permeability (Kun et al. 2014).

11.5.11 Self-Emulsifying Tablets

Self-emulsifying drug delivery system (SEDDs) is a most versatile drug delivery system used to prepare generally with the oil phase, aqueous phase, surfactant and co-surfactant(s). The optimized batch of SEDDs was dried using the freeze-drying method and then compressed into the tablet and investigated the in vitro drug release behavior and weight loss potential using the Wistar rats' model. The optimized SEDDS exhibited globule size 96 nm and emulsification time was 26 s. FTIPR analyze revealed that the drug and excipients were more compatible with each other. SEDDs-tablet was exhibited with rapid drug release behavior and elevated body weight reduction in rats as compared with the conventional tablet. Enhanced drug release profile and remarkable weight loss were achieved by the SEDDs-tablet approach entail that this delivery system could be the potential drug delivery concept with orlistat to control body weight (Gade and Hurkadale 2016).

11.5.12 Localized Delivery of Hydrogel

Metformin hydrochloride is an attachment of two guanidine structure to form a biguanide derivative which is popularly used for the treatment of type 2 diabetes mellitus around the world. Exploration of novel particulate carrier delivery system (e.g. nanoparticles, microparticles and nanocarriers) to metformin might potentially help to enhance the oral therapeutic efficacy and minimize the frequency of dosage administration, reduce the gastro-intestinal side-effects and drug-induced toxicity (Meltem and Selma 2016).

Ahmed and Hesham 2016, formulated to demonstrate the efficacy of locally administered metformin to the abdominal region to minimize intra-abdominal visceral fats than oral METHCL tablets. For obese people, high body weight is highly linked with intra-abdominal visceral and subcutaneous fat.

Though, METHCl tablets shown significant weight reduction in diabetic patients but it needs a higher dose which still poses a major challenge to the patients under the stress of METHCl side-effects. The investigation exhibited that the formulated METHCl loaded gel significantly reduces the abdominal size as compared with the placebo and METHCl tablets treated group. Moreover, METHCl tablets treated groups shown that it could not able to reduce the intra-abdominal fat in around half of the population. Thus, METHCl loaded hydrogel delivery system could be proposed as a potential drug delivery system to eliminate the deposited fatty compounds in the abdominal region (Ahmed and Hesham 2016).

11.5.13 Thermosensitive Hydrogel

The thermosensitive hydrogel system is incredibly essential biomaterials and has been investigated exponentially due to the sustained and controlled release behavior of therapeutic agents that is sensitive to modify in temperature. Leptin, a hormone that is secreted by adipose tissue and it is capable of effectively inhibiting food-intake and control body weight. Many research investigations on leptin have been executed during the early period for the treatment of obesity. Leptin is a major factor in the modulation of body weight, but suffers from poor therapeutic effect and side effects after systemic administration due to its shorter half-life and insignificant oral bioavailability, and administered frequently by injection, thus posing a major physical burden on the patient and associated administrative costs. Therefore, currently, there is a great demand of interest to prepare sustained or controlled release dosage forms to eliminate and overcome the aforementioned obstacles (US6541033).

The current invention explores a thermosensitive hydrogel delivery system for the parenteral administration of a therapeutic agent with biodegradable polymeric matrix to an animal model wherein a *in situ* gel depot is produced at the site of administration and the therapeutic agent is released either sustained/controlled release rate from the depot gel. Exendin-4 is a peptide that is obtained from Helodermatidae venom and exhibits a structural relationship to hormone glucagon-like peptide-1 (GLP-1) (Göke et al. 1993). Usually, protein and peptide compounds have a wide range of advantages including high aqueous solubility, excellent therapeutic activity, specificity and less toxicity. Though, their long list of applications is hampered due to physicochemical instabilities, including low pH value, enzymatic denaturation in the gastrointestinal region and swift elimination from the body (Ensign et al. 2012; Sung et al. 2012).

11.5.14 Nanoemulsion

Nanoemulsions are the nano-structured and kinetically stable colloidal delivery system. It has unique functional characteristics as compared with traditional emulsion. Nanoemulsions are non-irritant and non-toxic that can be utilized for a wide range of applications including topical or body cavities, and parenteral delivery (Jamuna and Ravishankar 2019).

Oleoresin capsicum (OC) is an oily resinous substance obtained from dried ripe red pepper (*Capcisum annuum* L) (Haas et al. 1997), which commonly acts as preservatives, taste enhancer and food additive. The capsicum extracts possess numerous pharmacological benefits including anti-obesity, anti-cancerous, anti-mutagenic, anti-oxidant, anti-microbial and anti-inflammatory activities (Zimmer et al. 2012) and anti-cancer activity (Dwivedi et al. 2011; Jeon et al. 2010). Interestingly, the transdermal delivery of capsaicin has gained great attention in therapeutic fields because of the powerful anti-inflammatory and antinociceptive properties of capsaicin (Joo-Yeon et al. 2014).

Joo-Yeon et al. (2014) analyzed the anti-obesity efficacy of nanoemulsion based oleoresin capsicum (N-OC) in high-fat diet-induced rats. From this investigation, N-OC formulation drastically inhibits the body weight with decreased adipose tissue and ameliorated plasma triglyceride levels in rats as compared with traditional OC formulation. Hence, the nanoemulsion system could serve as a potential carrier for the treatment of obesity (Joo-Yeon et al. 2014).

Conjugated linoleic acid (CLA) is isomers of linoleic acid with conjugated double bonds which are mainly found in milk, dairy products, and meat. CLA has drawn great importance for the last 4 decades in the nutraceutical field. Initially, CLA has discovered as an anti-cancerous agent and later explored to many biological applications including atherosclerosis, reduction of body fat and modulate inflammatory and immune function (Joo-Yeon et al. 2014). In recent days, CLA has fascinated interest as a weight-reducing compound due to its shown anti-obesity activity in animal and human studies through induction of apoptosis in adipocytes and decreased lipid accumulation in adipocytes through inhibition of lipoprotein lipase activity (Fiona Moloney et al. 2004). However, the therapeutic application of CLA has hindered due to its low aqueous solubility poor absorption and instability (Kristina Martinez et al. 2010). Hence, the nanoemulsion (NE) approach was employed to encapsulate with conjugated linoleic acid (NE-CLA) to enhance oral bioavailability. The anti-obesity efficacy of NE-CLA was investigated and the results demonstrated that NE conjugated CLA has potentially enhanced anti-obesity activity.

11.5.15 Polymeric In Situ Hydrogel Implants

Formulation of epigallocatechin gallate (EGCG) was prepared as polymeric *in situ* hydrogel implants using polylactide glycolic acid (PLGA). The formulated implants were analyzed for physical properties, drug content, in vitro drug release behavior and anti-obesity activity in high-fat diet-induced mice. The formulated *in situ* gel-based implants exhibited satisfactory physicochemical properties including clarity, pH and drug content). The *in situ* gels showed 89.67, 99.90, 89.0 and 87.02% drug release for the formulation E1, E2, E3 and E4, respectively at the end of 30 days. The selected implant E2 (containing EGCG and PLGA in 1:10) was administered subcutaneously (dose of 100 mg/kg) to the obese mice for 30 days period. After 30 days, the standard group (treated atorvastatin orally) shown a total of 47% reduction in body weight gain while the implant is shown 35% reduction in body weight gain and the hydrogel implant treated group showed decreases in the serum total cholesterol, triglyceride, higher values of serum HDL cholesterol and LDL-Cholesterol. The HDL/LDL ratio was found to be considerably improved with the implant treated mice (0.71 mg/dL) comparable to the standard group (0.91 mg/dL) against the group on a high-fat diet (0.09 mg/dL). This research finding suggested that the prepared in situ hydrogel polymeric implant can be used for an anti-hyperlipidemic and anti-obesity activity to achieve prolonged drug delivery (Zhang et al. 2015).

11.6 Future Challenges

Refined sophistication in the state of the art of technology on obesity formulations should be able to target desired tissue and deliver a sustained or controlled release. Consequently, many more research investigations to be explored to confirm the effectiveness of the nano-drug delivery approach to treat obesity with enhanced therapeutic efficacy with minimal or less toxicity.

References

Ahmed AA, Hesham MT (2016) Metformin loaded carbopol gel for lowering the intra-abdominal visceral fat. J Bioequiv Availab 8: 149-152

Araújo F, Shrestha N, Gomes MJ, Herranz-Blanco B, Liu D, Hirvonen JJ (2016) *In vivo* dual-delivery of glucagon like peptide-1 (GLP-1) and dipeptidyl peptidase-4 (DPP4) inhibitor through composites prepared by microfluidics for diabetes therapy. Nanoscale 8(20): 10706-10713

Balada F, Sanchis D, Grasa MM, Virgili J, Estruch J, Fernández-López JA, Remesar X, Alemany M (1997) Effect of the slimming agent oleoyl-estrone in liposomes on the body weight of Zucker obese rats. Int J Obes Relat Metab Disord 21(9): 789-795

Banks WA, DiPalma CR, Farrell CL (1999) Impaired transport of leptin across the blood-brain barrier in obesity. Peptides 20(11): 1341-1345

Barnhart KF (2011) A peptidomimetic targeting white fat causes weight loss and improved insulin resistance in obese monkeys. Sci Transl Med 3(108): 108-112

Berbari NF, Pasek RC, Malarkey EB (2013) Leptin resistance is a secondary consequence of the obesity in ciliopathy mutant mice. Proceedings of the National Academy of Sciences of the United States of America 110(19): 7796-7801

Bin F, Ping J, Yaohui N, Thomas K, Dale J, Nico VR, Zaiqing Y, Haiyan Xu (2011) Clodronate liposomes improve metabolic profile and reduce visceral adipose macrophage content in diet-induced obese mice. PLoS One 6(9): e24358

Boer AG, Gaillard PJ (2007) Drug targeting to the brain. Ann Rev Pharmacol Toxicol 47: 323-355

Chandrasekaran CV, Vijayalakshmi MA, Prakash K, Bansal VS, Meenakshi J, Amit A (2012) Herbal approach for obesity management. Am J Plant Sci 3: 1003-1014

Chawla A, Boisvert WA, Lee CH, Laffitte BA, Barak Y, Joseph SB, Liao D, Nagy L, Edwards PA, Curtiss LK (2001) A PPAR gamma-LXR-ABCA1 pathway in macrophages is involved in cholesterol efflux and atherogenesis. Mol Cell 7: 161-171

Cho EJ, Rahman MA, Kim SW, Baek YM, Hwang HJ, Oh JY, Hwang HS, Lee SH, Yun JW (2008) Chitosan oligosaccharides inhibit adipogenesis in 3T3-L1 adipocytes. J Microbiol Biotechnol 18: 80-87

Chunhui J, Liangju K, Madeline PM, Feng Y, Mario AC, Naagarajan N, Shihuan K, Meng D (2015) Biodegradable polymeric microsphere-based drug delivery for inductive browning of fat. Front Endocrinol (Lausanne) 6: 169

Colman E (2012) The FDA's assessment of two drugs for chronic weight management. N Engl J Med 367(17): 1577-1579

Cooke D, Bloom S (2006) The obesity pipeline: current strategies in the development of anti-obesity drugs. Nat Rev Drug Discov 5: 919-931

Dowling RC, Grobert (2011) Nanoscience and nanotechnologies: opportunities and uncertainties. The Royal Society & The Royal Academy of Engineering Report, Royal Academy of Engineering, London, UK, 2004

Dwivedi V, Shrivastava R, Hussain S, Ganguly C, Bharadwaj M (2011) Cytotoxic potential of Indian spices (extracts) against esophageal squamorimona carcinoma cells. Asian Pac J Cancer Prev 12(8): 2069-2073

Ensign LM, Cone R, Hanes J (2012) Oral drug delivery with polymeric nanoparticles: the gastrointestinal mucus barriers. Adv Drug Deliver Rev 64: 557-570

Fathima A, Saravana Kumar P, Ramya Devi D, Vedha Hari BN (2012) Microparticulate drug delivery system of orlistat for sustained release. Int J Pharm Sci Rev Res 15(1): 27-30

Fiona Moloney, Toh-Peng Yeow, Anne Mullen, John J Nolan, Helen M Roche (2004) Conjugated linoleic acid supplementation, insulin sensitivity, and lipoprotein metabolism in patients with type 2 diabetes mellitus. Am J Clin Nutr 80(4): 887-895

Food and Drug Administration (FDA), Guidance for Industry: Assessing the Effects of Significant Manufacturing Process Changes, Including Emerging Technologies, on the Safety and Regulatory Status of Food Ingredients and Food Contact Substances, Including Food Ingredients That are Color Additives, Food and Drug Administration (FDA), Silver Spring, Md, USA, 2012

Friedman JM (2009) Obesity: Causes and control of excess body fat. Nature 459(7245): 340-342

Gabathuler R (2010) Approaches to transport therapeutic drugs across the blood-brain barrier to treat brain diseases. Neurobiol Dis 37: 48-57

Gade MM, Hurkadale PJ (2016). Formulation and evaluation of self-emulsifying orlistat tablet to enhance drug release and *in vivo* performance: factorial design approach. Drug Deliv Transl Res 6(3): 276-288

Göke R, Fehmann HC, Linn T, Schmidt H, Krause M (1993) Exendin-4 is a high potency agonist and truncated exendin-(9-39)-amide an antagonist at the glucagon-like peptide 1-(7-36)-amide receptor of Insulin-secreting β-cells. J Biol Chem 268: 19650-19656

Gupta BK, Manas Chakraborty, Debnath R, Soumyadip B, Nandy S (2010) Design and evaluation of rimonabant loaded sodium alginate-guar-gum microsphere formulation. IJPSR 1(8): 47-61

Haas JS, Whipple RE, Grant PM, Andresen BD, Volpe AM, Pelkey GE (1997) Chemical and elemental comparison of two formulations of oleoresin capsicum. Sci Justice 37(1): 15-24

Holst J (2007) The physiology of glucagon-like peptide 1. Physiol Rev 87: 1409-1439

Hossen MN, Kajimoto K, Akita H, Hyodo M, Harashima H (2012) Vascular-targeted nanotherapy for obesity: unexpected passive targeting mechanism to obese fat for the enhancement of active drug delivery. J Control Release 28(2): 101-110. http://www.google.com.br/patents/US6541033; accessed on 16th September 2019

Dowman JK, Tomlinson JW, Newsome PN (2010) Pathogenesis of non-alcoholic fatty liver disease. QJM 103(2): 71-83

Jamuna BA, Ravishankar RV (2019) Nanoemulsions and their potential applications in food industry. Front Sustain Food Syst 13(3): 1-21

Jeon G, Choi Y, Lee SM, Kim Y, Jeong HS, Lee J (2010) Anti-obesity activity of methanol extract from hot pepper (*Capsicum annuum L.*) seeds in 3T3-L1 adipocyte. Food Sci Biotechnol 19(4): 1123-1127

Joo-Yeon K, Mak-Soon L, Sunyoon J, Hyunjin J, Chong-Tai K, In-Hwan K, Sangjin S, Soojung O, Yangha K (2014) Anti-obesity efficacy of nanoemulsion oleoresin capsicum in obese rats fed a high-fat diet. Int J Nanomedicine 9: 301-310

Joseph R, Elliot LF (1997) LIPOSOME FORMULATIONS OF 5β STEROIDS WO 1997013500 A2 Application no: PCT/US1996/015507

Jyotibala B (2017) Application of nanotechnology in food technology and targeted drug therapy for prevention of obesity: An overview. J Crit Rev 4(1): 7-11

Kristina Martinez, Arion Kennedy, Tiffany West, Dejan Milatovic, Michael Aschner, Michael McIntosh (2010) J Biol Chem 285 (23): 17701-17712

Kumar SG, Rahman MA, Lee SH, Hwang HS, Kim H, Yun JW (2009) Plasma proteome analysis for anti-obesity and anti-diabetic potentials of chitosan oligosaccharides in ob/ob mice. Proteomics 9: 2149-2162

Kumari A, Yadav SK, Yadav SC (2010) Biodegradable polymeric nanoparticles based drug delivery systems. Colloids and surfaces B: Biointerfaces 75(1): 1-8

Kun L, Shuanshuan X, Shilong H, Jidong Z, Xinwen C, Jin C, Qingqing H, Qing Y, Haiyan L, Lei X (2014) Preparation of a nano emodin transfersome and study on its anti-obesity mechanism in adipose tissue of diet-induced obese rats. J Transl Med 12: 72

Lanlan H, Jian C, Peiqiu C, Haitao P, Chen D, Tiancun X, Pengfei Z, Jiao G, Zhengquan S (2015) Anti-obese effect of glucosamine and chitosan oligosaccharide in high-fat diet-induced obese rats. Mar Drugs 13: 2732-2756

Lichota J, Skjorringe T, Thomsen LB, Moos T (2010) Macromolecular drug transport into the brain using targeted therapy. J Neurochem 113: 1-13

Khazaei M, Tahergorabi Z (2013) Systemic ghrelin administration alters serum biomarkers of angiogenesis in diet-induced obese mice Int J Pept ID 249565: 5

Madureira AR, Pereira A, Pintado M (2015) Current state on the development of nanoparticles for use against bacterial gastrointestinal pathogens. Focus on chitosan nanoparticles loaded with phenolic compounds. Carbohydr Polym 130: 429-439

Margalit R, Yerushalmi N (2005) Pharmaceutical aspects of liposomes: academic and industrial research and development. Microencapsulation Methods and Industrial Applications 1: 317-344

Meltem C, Selma S (2016) Microparticulate and nanoparticulate drug delivery systems for metformin hydrochloride. Drug Deliv 23(8): 2796-2805

National Center for Health Statistics, Hyattsville, Md (1999) Prevalence of overweight or obesity among adults: United States. Health E-Stats 2000

Nazarii K, Ludovico A, Tetyana F, Oleksandr V, Belemets N, Tetyana B, Petro B, Mykola S (2016) Prevention of NAFLD development in rats with obesity via the improvement of pro/antioxidant state by cerium dioxide nanoparticles. Clujul Med 89(2): 229-235

Overweight, obesity, and health risk (2000) National task force on the prevention and treatment of obesity. Arch Intern Med 160: 898-904

Padwal RS, Majumdar SR (2007) Drug treatments for obesity: orlistat, sibutramine, and rimonabant. Lancet 369: 71-77

Rahman A, Kumar SG, Kim SW, Hwang HJ, Baek YM, Lee SH, Hwang HS, Shon YH, Nam KS, Yun JW (2008) Proteomic analysis for inhibitory effect of chitosan oligosaccharides on 3T3-L1 adipocyte differentiation. Proteomics 8: 569-581

Reinbach, H C, Smeets A, Martinussen T, Møller P, Westerterp-Plantenga, MS (2009) Effects of capsaicin, green tea and CH-19 sweet pepper on appetite and energy intake in humans in negative and positive energy balance. Clinical Nutrition 28(3): 260-265

Sargent, BJ, Moore NA (2009) New central targets for the treatment of obesity. Br J Clin Pharmacol 68: 852-860

Shikha J, Vikas J, Mahajan SC (2014) Lipid based vesicular drug delivery systems. Adv Pharm Article ID 574673: 12

Singer K, Morris DL, Oatmen KE (2013) Neuropeptide Y is produced by adipose tissue macrophages and regulates obesity-induced inflammation. PLoS One 8(3): e57929

Sung HW, Sonaje K, Liao ZX, Hsu LW, Chuang EY (2012) pH-responsive nanoparticles shelled with chitosan for oral delivery of insulin: from mechanism to therapeutic applications. Acc Chem Res 45: 619-629

Sunil KJ, Govind PA, Narendra KJ (2006) Evaluation of porous carrier-based floating orlistat microspheres for gastric delivery AAPS Pharm Sci Tech 7(4): E54-E62

Tan S, Gao B, Tao Y, Guo J, Su ZQ (2014) Anti-obese effects of capsaicin-chitosan microsphere (CCMS) in obese rats induced by high fat diet. J Agric Food Chem 62(8): 1866-1874

Tiwari SB, Amiji MM (2006) A review of nanocarrier-based CNS delivery systems. Curr Drug Deliv 3: 219-232

Toita R, Kawano T, Murata M, Kang JH (2016) Anti-obesity and anti-inflammatory effects of macrophage-targeted interleukin-10-conjugated liposomes in obese mice. Biomaterials 110: 81-88

Thovhogi N, Sibuyi N, Meyer M, Onani M, Madiehe A (2015) Targeted delivery using peptide-functionalised gold nanoparticles to white adipose tissues of obese rats. J Nanopart Res 17(2): 112

Tosi G, Bortot B, Ruozi B, (2013) Potential use of polymeric nanoparticles for drug delivery across the blood-brain barrier. Curr Med Chem 20: 2212-2225

Tosi G, Costantino L, Rivasi F (2007) Targeting the central nervous system: *in vivo* experiments with peptide-derivatized nanoparticles loaded with loperamide and rhodamine-123. J Control Release 122: 1-9

Tuerk C, Gold L (1990) Systemic evolution of ligands by exponential enrichment: RNA ligands to bacteriophage T4 DNA polymerage. Science 249: 505-510

US 20150190531 A1 Nanoparticle delivery system for targeted anti-obesity treatment

US Department of Health and Human Services. The Surgeon General's Call to Action to Prevent and Decrease Overweight and Obesity. Washington, DC, 2004

Van Heek M, Compton DS, France CF, Tedesco RP, Fawzi AB, Graziano MP, Sybertz EJ., Strader CD, Davis HR (1997) Diet-induced obese mice develop peripheral, but not central, resistance to leptin. J Clin Investig 99: 385-390

Zhang XZ, Guan J, Cai SL, Du Q, Guo ML (2015) Polymeric *in situ* hydrogel implant of epigallocatechin gallate (EGCG) for prolonged and improved antihyperlipidemic and anti-obesity activity: preparation and characterization. J Biomater Tissue Eng 5(10): 813-817

Zimmer AR, Leonardi B, Miron D, Schapoval E, Oliveira JR, Gosmann G (2012) Antioxidant and anti-inflammatory properties of *Capsicum baccatum*: from traditional use to scientific approach. J Ethnopharmacol 139(1): 228-233

Computational Approach in Drug Development for Obesity

12

Asita Elengoe

Abstract

Currently, obesity is a major menacing global health problem and the leading cause of mortality among men and women. The incidence of obesity and overweight has been escalating. Malaysian statistics showed that there was a great increase in the past two decades. The increase is attributed to a lack of awareness of obesity and its risk-related diseases. Currently, the available treatments such as drugs and surgery have not produced significant results to overcome the obesity problem. Therefore, the development of a novel therapeutic approach to obesity remains one of the most challenging medical problems. Phyto-compounds have been identified as a class of promising anti-obesity agents with fewer or no side effects in humans. Substrate-based drug design (SBDD) is an effective computational biology method to find the anti-obesity drug. Thus, the new target protein is constructed using Swiss model and molecular dynamics simulated. Then, it was validated by PROCHECK, ERRAT, WHATCHECK, WHATIF, PROSA, Verify 3D, Pro-Q, GROMOS and ANOLEA. The active sites of the target protein are obtained using Q-site Finder or Active site prediction server. After that, molecular docking is performed to determine the binding affinity between the target protein and plant compounds. The best docking score is determined based on the lowest binding energy (kcal/mol) and the highest number of hydrogen bonds in their interaction. Finally, the stability target protein complex compound is assessed by molecular dynamics (MD) simulation tools. Therefore, the phyto-compound structure-based drug discovery may be potential as an anti-obesity agent.

Keywords

Obesity · Target protein · Phyto-compound · Computational tools

A. Elengoe (✉)
Biotechnology Department, Faculty of Science, Lincoln University College, Petaling Jaya, Selangor, Malaysia

M. K. Kutty, A. Elengoe (eds.), *Obesity and its Impact on Health*,
https://doi.org/10.1007/978-981-33-6408-0_12

12.1 Introduction

Obese subjects have a body mass index (BMI) of 30 or higher. Obesity is defined as the health condition of anyone significantly above his or her ideal healthy weight. Currently, obesity is a major global health issue problem and the leading cause of morbidity and mortality among men and women. Trends of obesity and overweight were increasing from 1996 to 2015 (World Health Organization 2015). According to Malaysian statistics, there was a great increase in the last 20 years (Khambalia and Seen 2010). The prevalence of obesity has increased significantly from 2006 (43.1%), 2011 (44.5%) and 2015 (47.7%) (Alwan et al. 2011). Based on National Health and Morbidity Survey (NHMS) (2015), it was approximated that 5.6 million and 3.3 million Malaysian adults (aged >18) were overweight and obese respectively. In 2013, 4.4 million Malaysian children suffered from obesity. Mostly, people who are aged between 40 and 60 years old suffered from obesity. Moreover, nearly half of those afflicted are 60 years old. This is because of unhealthy diet (fast food such as KFC, McDonald, Burger King, Domino, Pizza Hut, etc); lack of physical activity; genetic disorders; and lack of awareness on obesity and its risk-related diseases such as diabetes mellitus (DM) type 2, heart-related diseases, hypertension, polycystic ovarian syndrome (PCOS), endometriosis, respiratory problems, kidney-related diseases, osteoarthritis, etc. At present, the available treatments such as pharmaceutical drugs and surgery have not been fully effective against the rising incidence of obesity. Furthermore, these modern treatment modalities cause adverse effects such as kidney, liver failures and developing other types of health complications.

Therefore, finding a solution to obesity remains one of the most challenging area in the medical field.

Computational biology is one of the most promising tools for the discovery and design of phyto-compound drugs for obesity. At present, this strategy is growing rapidly due to low cost, save energy and time compared to laboratory experiments. Plant compounds have been recognized as a class of potential anti-obesity agents in the search for novel pharmaceutical compounds as they can produce secondary metabolites with new chemical structures. Phyto-compounds are safe and less toxic to normal cells compared to chemical drugs.

12.2 Obesity-Related Comorbid Diseases

Obesity escalates the risk of developing several comorbid diseases. It induces diabetes type 2 disease. The hormone insulin is actively involved in the transport of blood glucose into the muscle and fat cells. This negative regulation maintains blood glucose levels in normal conditions. However, insulin resistance means that blockage of transport of blood glucose into the cells. Fat cells are more resistant than muscle cells. Thus, obesity is the main cause of insulin resistance. Initially, the pancreas responds to insulin resistance by stimulating more insulin. The pancreas struggles for a longer time to produce sufficient insulin to overcome this insulin resistance so that, it can maintain the normal blood glucose level. This insulin resistance state can last for years. However, once the pancreas fails to maintain the normal range of blood glucose by generating increase levels of insulin, resulting in diabetes mellitus type 2. Thus, insulin resistance represents a pre-diabetes status. The risk of diabetes mellitus type 2 depends on the duration and degree of obesity.

Besides insulin resistance, hypertension (high blood pressure) is a common disease among obese adults. Previous studies have proven that the gain of weight tended to raise the blood pressure high significantly in women compared with men. Obesity also can enhance blood cholesterol (hypercholesterolemia) and triglycerides level in the body. It reduces the "good" high-density lipoprotein (HDL) cholesterol. HDL cholesterol destroys bad and harmful cholesterol. Thus, a high level of HDL decreases the risk of heart disease and stroke (cerebrovascular accident). However, a low level of HDL raises the tendency of risk.

Furthermore, obesity tends to escalate the risk of a heart attack. Clinical studies demonstrated that the increased risk of developing coronary artery disease three to fourfolds in women whose BMI exceeds 29. A Finnish study described that every 1 kg of weight gain augments the mortality risk from coronary artery disease by 1%.

Obesity also tends to develop different types of cancer such as colon, rectum, prostate, gallbladder, uterus and breast cancer among men and women. Adipose tissue plays an important role in the production of the hormone estrogen. Yet, prolonged exposure to high levels of estrogen enhances the risk of breast cancer.

12.3 Etiology of Obesity

Numerous factors cause the development of obesity. However, overconsumption of food and physical inactivity are the most common causes of obesity. Bodyweight is the result of genetic defects, metabolism, behaviour, culture and environment.

Overeating induces an increase in weight; especially if the food is high in fat. Food consists of high fat or sugar such as fast food (KFC, McDonald, Burger King and Pizza Hut), fried food (French fries, hotdog, and nuggets) and sweets have energy density. Even a small quantity of food has high calories. Studies have established that diets high in fat lead to weight gain. According to some researchers, simple carbohydrates such as sugars, fructose, desserts, soft drinks, beer, chocolates, potato chips lead to weight gain because they are absorbed into the blood circulation very quickly compared to complex carbohydrates like pasta, spaghetti, rice, vegetables, fruits. Hence, higher levels of hormone insulin are required for the metabolism of simple carbohydrates than complex carbohydrates. The elevated hormone insulin release induces weight gain.

Besides, studies demonstrated that people who consume a small amount of food 4 or 5 times daily have lower cholesterol levels and stable blood sugar levels compared to those who eat a large amount of food 2 or 3 times daily. This is because the pancreas releases a low level of hormone insulin for small frequent meals, while the high amount of insulin is required to break down the large meal.

Furthermore, people who are active burn more calories than inactive people. It has been proven that physical inactivity is strongly associated with weight gain in both genders (men and women). Regular exercise reduces the fats in the body.

Medications such as antidepressants, anticonvulsants (carbamazepine and valproate), diabetes (insulin, sulfonylureas and thiazolidinediones), certain hormones (oral conceptive and corticosteroids), high blood pressure medications and antihistamines can be responsible for weight gain. It is of interest that psychological factors also contribute to obesity. Emotions such as anger, boredom, happiness, sadness, stress or anger influence eating habits. Besides, hypothyroidism, insulin resistance, polycystic ovary syndrome (PCOS), and Cushing's syndrome diseases also cause obesity.

Moreover, additional factors such as inadequate monitory circumstances to buy healthy food or lack of safe places for jogging and walking can increase the risk of obesity. Imbalance of hormones also leads to weight gain, especially during pregnancy and menopause.

12.4 Genetic Alterations of Obesity

If one or both parents are obese, the person has a high chance to get obesity problems. Genetic alteration can affect the hormones implicated in the mechanism of fat regulation. Hormone leptin deficiency is a genetic defect associated with obesity. Leptin is produced in fat cells and the placenta. It controls the weight by sending signals to the hypothalamus (at brain) to eat less when the body fat is very high. The circumstances leading to insufficient leptin or failure of leptin to transmit a signal to the brain to eat less, the mechanism is disrupted with resultant obesity. Currently, the role of leptin replacement therapy for obesity is being explored.

12.5 Roles of Medicinal Plants in Obesity

Phyto-compounds are a class of promising anti-obesity agents. Their potential as anti-obesity agents has been investigated intensively in the treatment of obesity. Moreover, numerous ethnobotanical studies have promoted the use of herbal medicine instead of synthetic drug treatment for obesity. The advantages of herbal therapy are (1) its relative safety, (2) absence of adverse effects and (3) effective therapeutic effect. Ethanopharmacology aids to identify the novel biologically active compounds from plant sources. Eg. Plant compounds from *Citrus limon* (lemon), *Allium sativum* (garlic), *Zingiber officinale* (ginger), *Psidium guajava* (guava), *Camellia sinensis* (tea), *Curcuma longa* (turmeric) and *Nigella sativa* (black cumin) have the potential to be anti-obesity drugs. The computational biology tool has the advantage to find and design anti-obesity drugs via discovering the interaction between the target protein and ligand (plant compounds). The uses of medicinal plants in obesity are quoted from the previous research findings and summarized in Table 12.1.

Table 12.1 Mechanisms of the bioactive compounds of medicinal plants

No	Bioactive compounds	Mechanism	Reference
1	Flavan-3-ol,catequins	-Increases the expression of gene that related to energy expenditure on skeletal muscle -Reduces the synthesis of fat	Ikarashi et al. (2011)
2	Phenols, flavonoids, saponins	-Inhibits enzyme activity of the pancreatic amylase and lipase to delay the intestinal absorption of fat in the diet	Dey (2011)
3	Saponins, β-asarone	-Inhibits adipogenesis and stimulates lipolysis in 3T3-L1 adipocytes	Singh et al. (2011)
4	Flavonoids	-Reduces the triglycerides and cholesterol levels in blood through inhibition of the enzyme lipase	Dickel et al. (2007)
5	Cumarins	-Reduces triglyceride and cholesterol levels of lipid metabolism modulation	Karmase et al. (2013)
6	Phenols	-Inhibits adipogenesis in3T3-1 pre-adipocyte cells	Sung et al. (2015)
7	Phenols	-Reduces cholesterol and other blood lipids in blood -Lowers the activity of genes and enzymes that facilitate deposition and formation of fats and glucose	Tanaka et al. (2006)
8	Diallylsulfide (DAS), Diallyldisulfide (DADS) or Diallyltrisulfide (DATS)	-Inhibits differentiation of adipocytes	Bordia et al. (1998)
9	Alkaloids, terpenes	-Reduces lipid levels -Increase high density lipoprotein-C (HDL-C) levels	Cercato et al. (2015)
10	Tannins, alkaloids	-Suppresses the formation of lipid droplets in adipocytes -Reduces false hunger (pseudo hunger)	Bhat et al. (2017)
11	Phenols, terpenes	-Suppresses lipogenic enzymatic activities -Reduces triglycerides accumulation in liver and adipose tissues	Kim et al. (2010)
12	Pentacyclic triterpenoids	-Increase fat metabolism -Reduces cholesterol and triglycerides level in blood	Alam et al. (2012)
13	Phenols, flavanoids, alkaloids	-Reduces blood levels of triglycerides and non-esterified fatty acids	Moro and Basile (2000)
14	Cathinone	-Suppress hunger and increase endurance	Kamalakkannan et al. (2010)
15	Saponins,alkaloids, tannins, anthraquinones, flavonoids	-Suppress appetite stimulus signals in the hypothalamus -Reduces triglycerides, low density	Krishna et al. (2008)

(continued)

Table 12.1 (continued)

No	Bioactive compounds	Mechanism	Reference
		lipoprotein –C (LDL-C) and very low density lipoprotein-C (VLDL-C) levels in blood	
16	Catechins,epicatechin	-Inhibits activity of lipoprotein, LPL	Brinkhaus et al. (2000)
17	Procyanidins, catechins	-Increase lipolysis through inducing β-oxidation or noradrenaline secretion in fat cells	Yuliana et al. (2011)
18	Saponins	-Suppress triglycerides level in the blood	Kwon et al. (2003)
19	Tannins	-Inhibits differentiation of adipocytes	Vasudeva et al. (2012)
20	Flavanoids, Phenol	-Enhances beta oxidation through upregulation of mRNA level in liver and white adipose tissue	González-Molina et al. (2010)
21	Terpenes	-Reduces triglycerides and cholesterol levels of lipid metabolism	Choi et al. (2007)
22	Curcumin, Demethoxycurcumin, bisdemethoxycurcumin	-Inhibits adipogenesis	Nishiyama et al. (2005)
23	Quercitol, alkaloids	-Altering insulin resistance and insulin signaling in hepatocytes, which consequently promote increased intracellular fatty acid	Nazish et al. (2012)
24	Triterpenoids	-Reduces accumulation of triglycerides in 3T3-L1 adipocytes	Seto et al. (2009)
25	Hydroxycitric acid	-Reduces low density lipoprotein and cholesterol levels in plasma and increases chylomicrons level - Suppresses enzyme adenosine triphosphatase (ATPase) citratelyase to block lipogenesis activity	Heymsfield et al. (1998)
26	Flavonoids, sterols, amino acids, chalcones, triterpenoid, saponins	-Reduces blood levels of cholesterol and triglycerides	Nakagawa et al. (2004)
27	Alkaloids, triterpenes, flavonoids, steroids, saponins	-Inhibits the accumulation of triglycerides in muscle and liver -Decrease fatty acid accumulation in the blood circulation	Pothuraju et al. (2014)
28	Saponins,flavanoids	-Suppress the consumption of sugar -Reduce blood levels of cholesterol and triglycerides	Austin and Herbals (2008)
29	Phenols, flavonoids, anthocyanins	-Inhibits the adipocyte differentiation through a modulation of the P13-K and MAPK pathways	Alarcon-Aguilar et al. (2007)

(continued)

Table 12.1 (continued)

No	Bioactive compounds	Mechanism	Reference
30	Alkaloids	-Inhibits intestinal absorption and biosynthesis of lipid	Adeneye et al. (2010)
31	Methylxanthines, saponins, alkaloids	-Reduces triglyceride and cholesterol levels of lipid metabolism -Increase the expression of uncoupling proteins and raised AMPK phosphorylation in visceral adipose tissue	Bracesco et al. (2011)
32	Tannins	-Inhibits differentiation of adipocytes	Klein et al. (2007)
33	Catechins, epicatechin	-Inhibits the activity of lipoprotein lipase, LPL	Nerurkar et al. (2010)
34	Catechins,epicatechin	-Inhibits the activity of lipoprotein lipase, LPL	Governa et al. (2018)
35	Fitotesrol (β-sitosterol)	-Reduces atherogenic index, reversing the hyperlipidemic effect produced by high fat diet	Falowo et al. (2018)
36	Alkaloids	-Increase lipolytic effects with decreased intracellular triglycerides and release of glycerol	Munhoz and Frode (2018)
37	Tetrahydrofuran	-Reduces LDL and glucose level through adenosine monophosphate (AMP)-activated protein kinase activation mechanism	Rédei et al. (2018)
38	Alkaloids	-Suppress enzyme activity of the pancreatic amylase and lipase to delay the intestinal absorption of fat in the diet	Lee et al. (2015)
39	Carotenoids, terpenoids, flavonoids, polyphenols, alkaloids, tannins, saponins	-Reduce blood levels of the cholesterol and triglycerides	Sharma and Singh (2012)
40	Flavonoids	-Reduce blood levels of the cholesterol and triglycerides	Lee et al. (2011)
41	Triterpenoids	-Inhibits pancreatic lipases from hydrolyzing the ingested fat into absorbable free fatty acids and monoglycerides.	Chandrasekaran et al. (2012)
42	Phenolics, alkaloids, flavonoids,terpenoids	-Inhibits human lipoxygenase	Ahmad et al. (2016)
43	Mangiferin, catechins, condensed tannins	-Inhibits pancreatic lipase (LPL) and glycerophosphate dehydrogenase (GPDH)	Yoshikawa et al. (2002)
44	Phenols	-Reduces blood level of cholesterol	Audu et al. (2014)

(continued)

Table 12.1 (continued)

No	Bioactive compounds	Mechanism	Reference
45	Tannins, saponins, flavonoids, terpenoids, phenolics, steroids, and alkaloids	-Increase fat metabolism	Bunawan et al. (2015)
46	Polyphenols	-Decrease total cholesterol, low-density lipoprotein (LDL), and triglycerides in plasma -Increase high-density lipoprotein (HDL)	Azman et al. (2012)
47	Alkaloids, flavanoids	-Inhibits the absorption of lipid	Raju and Bird (2006)
48	Polyphenols	-Inhibits pancreatic lipase	Kim et al. (2014)
49	Phenols	-Decreases total lipids, cholesterol and triglycerides in blood	Manchanda and Kaur (2017)
50	Polyphenols, curcumin	-Increases thermogenesis -Increases lipolysis -Suppresses lipogenesis -Inhibits intestinal fat absorption and controlling appetite	Ali et al. (2008)

12.6 Drug Design Using Computational Biology Tool

Computational biology is an important and potential branch of bioinformatics where a computer system, is used to analyse and store biological data. At present, this approach is receiving considerable attention from fields such as medicine, agriculture, pharmaceutical and others. The advantages of this are (1) cost-effectiveness, (2) time saving and (3) reducing manpower compared to laboratory experiments. This application of advanced technology helps to discover the novel therapeutic interaction of target protein and ligand (plant compound) in the development of the drug for the obese population.

12.6.1 Structure-Based Drug Design

Currently, the application of structure-based drug design (SBDD) is possible and effective because they are routinely applied to medical and pharmaceutical research. Therapeutic agents for the diseases are being developed for specific drug designs. Drugs are designed based on the knowledge of the biological target and structural information provided through structural biology.

In the past century, the information of structure and biological compounds' target was unknown and hence, trial and error methods were performed to screen and optimize the biologically active compounds and chemicals (Macarron 2006). Currently, SBDD helps to focus on the target research and allows the discovery and design of the drug with high binding affinity and selectivity against specific targets (eg. proteins, carbohydrates, lipids, DNA, RNA, etc.). Moreover, with a complete comprehension of the three-dimensional (3-D) model of the targets (macromolecules), SBDD favors a great opportunity to target allosteric sites, thus expanding the possible pathway to regulate the targets' functions either through enhances or inhibits a particular activity (eg. transcription, translation, DNA binding, molecular chaperone activity, receptor signalling protein activity, signal transducer activity, protein kinase activity, promoter binding, etc). This technique investigates how the candidate drug (phytocomponent) interacts with the target protein by analysing their binding affinity.

12.6.2 Computational Biology

There are different types of protein modelling such as homology modelling or comparative protein modelling, threading, ab initio and de novo approaches. Currently, homology modelling is widely used in drug design because of the generation of high accuracy levels of the three-dimensional model (3-D) of the protein of interest. Generally, 30% sequence identity is required to build a perfect model.

There are four main steps involved in homology modelling. Firstly, the target protein sequence is obtained from NCBI in FASTA format. After that, BLASTP against the RCSB Protein Databank is carried out to identify a suitable template with the existing three dimensional (3-D) model of the template protein sequence. Then, the target protein sequence is aligned with the template sequence using multiple sequence alignment. CPHmodel, EsyPred3D and Swiss-model are used to generate the 3-D protein model based on given sequence alignment and selected template from Protein Data Bank (PDB). The experimental 3-D structures of proteins (templates) can be found in PDB. Next, the protein model is refined using molecular dynamics (MD) simulation tools such as GROMACS, NAMD, CHARMM, AMBER, DESMOND, TINKER, and DL_POLY. Lastly, the stability and quality of the generated protein model are evaluated using the protein model validation tools such as PROCHECK, ERRAT, WHATCHECK, WHATIF, PROSA, Verify 3D, Pro-Q, GROMOS and ANOLEA.

Then, the active sites of the target protein are identified either through Q-site Finder or Active Site Prediction Server (Jayaram 2004). The target protein will be docked with the plant compound using the Autodock tool. There are two types of AutoDock tools such as AutoDock 4 and AutoDock Vina (Sanner 1999). The most stable target protein and phyto-compound complex is evaluated by their binding affinities. Binding affinity depends on the lowest binding energy and the presence of the highest number of hydrogen bonds within the complex structure. Finally, MD simulation of target protein-phyto-compound complex is performed to assess their stability throughout the simulation period. The docked target protein and phyto-compound-complex structure is analysed.

12.6.3 Current Computational Approaches in Obesity

Akhila et al. (2020) studied the molecular docking between caffeic acid, coumaric acid, quercetin, genistein, daidzein, apigenin, luteolin, catechin, malvidin, capsaicin, nicotine, resveratrol (plant compounds); orlistat (synthetic drug) and obesity-associated proteins (Fat Mass and Obesity-associated protein (FTO), leptin and resistin). The plant compounds and synthetic drug were the ligands while obesity-associated proteins were target proteins. They found that daidzein could be the potential phyto-compound for the anti-obesity drug because it docked successfully with FTO, leptin and resistin at lowest binding energies (the most negative values) such as −6.84, −5.81 and −5.99 kcal/mol respectively.

Based on Zoy et al. (2019) study, linalool, estragole, β-elemene, caryophyllene, α-Bergamotene, γ-cadinene, phytol, squalene, heptacosane and hexatriacontane were screened for potential binding with the procine-pancreatic lipase (PPL). These plant compounds may have the potential to be used for anti-obesity treatment. Linalool had the best binding affinity (−6.5 kcal/mol) with PPL compared with estragole (−6.1 kcal/mol).

According to Alamri Mohammed et al. (2015) study, they investigated the binding affinity of exemestane, kaempherol, quercetin, letrozole and rutin (phyto-compounds) and orlistat against the obesity-associated protein (FTO). The results revealed that quercetin had the best binding affinity with FTO. The binding energy was—1.78 e^{+32} kcal/mol. Therefore, quercetin could be a potential anti-obesity drug. Many flavonoids have been demonstrated that they play an essential role in lipolysis and adipogenesis in adipose cells (Kuppusamy and Das 1992; Kuppusamy and Das 1994).

Based on Gandhi et al. (2019) study, niacin, quercetin 3, 7 diglucoside, ascorbic acid, quercetin 3, 3′ diglucoside, quercetin 3,4′ diglucoside, 8 nonynoic acid, 9 decynoic acid, cyanidine 3, 5 diglucoside, riboflavin, thiamine, beta rosasterol, cyanidin 3-sophoroside-5-glucoside, methyl non-8-ynoate, methyl dec-9-ynoate, methyl (e)-11-methoxy-9-oxononadec-10-enoate, methyl malvalate, methyl 8-oxooctadec-9-ynoate, methyl sterculate, campesterol, stigmasterol, beta sitosterol (phyto-compounds), orlistat (synthetic drug) were screened for potential binding with obesity-associated proteins such as pancreatic lipase (target). The results demonstrated that niacin could be the potential plant bioactive compound for the anti-obesity drug because it docked successfully with pancreatic lipase at the lowest binding score (−27.2868). They also found that riboflavin bound strongly with FTO, leptin and SCH1 at the most negative values such as −27.3248, −18.4869 and −13.553 respectively. While niacin docked successfully with cannabinoid receptor and ghrelin at −14.7132 and −11.1374 respectively. Moreover, quercetin, 3, 3′ diglucoside had a strong interaction with MCH1 protein. The binding score was −13.7266. Pancreatic lipase, FTO, cannabinoid receptor, ghrelin, quercetin, 3, 3′ diglucoside, SCH1 and MCH1 are the ligands. These phytochemicals of *Hibiscus rosa-sinensis* may have the potential to be used for anti-obesity treatment.

According to Sivashanmugam et al. (2013) study, they investigated the binding affinity of ten phytocomponents (apigenin, biochanin, curcumin, diosmetin, epipervilline, glycitein, okanin, rhamnazin, tangeritin and gallic acid) against cholesterol esterase (target). Gallic acid was used as a standard phytocomponent, thus it was compared with other plant compounds. The results revealed that the plant compounds had good interaction with cholesterol esterase due to negative binding energy values which ranged between −7.08 and −4.11 kcal/mol. This shows that these phytocomponents suppressed the cholesterol esterase activity and could be potential drugs in obesity treatment.

Alias et al. (2020) studied molecular docking between leucoanthocyadin, rutin and astragalin (plant bioactive components) and pancreatic lipase (human PL-colipase (CL) (PDB ID: 1LPB) using Autodock 4.2 tool. The plant components derived from the leaf extract of *Nelumbo nucifera*. They found that leucoanthocyadin, rutin and astragalin could be potential phyto-compounds for inhibition of human pancreatic lipase activity. However, astragalin had the strongest binding affinity with the target. SER 152, HIS 151, PHE 215, and ARG 256 are the most potent active sites for drug targets. They had strong hydrogen bond interactions with astragalin.

In silico investigation was carried out to identify the potential phytocompounds of *Punica granatum* (leaf extract) against pancreatic lipase and FTO by Sudeep and Prasad (2014). Orlistat was used as standard. The plant compounds were punicalagin, ellagic acid, corilagin, brevifolin, punicallin, apigenin and luteolin 4-O-glucopyranoside. They were docked successfully with pancreatic lipase and FTO using Autodock 4.2 tool. Punicalagin had the strongest bond with pancreatic lipase at the lowest binding energy (−11.85 kcal/mol). It interacted with the active residues of pancreatic lipase (ASN 384, TYR 369, ARG 337, ASN 229, ASP 387 and ARG 339). It had the highest number of hydrogen bonds (7) with pancreatic lipase. On the other hand, corilagin had the strongest binding affinity (−9.44 kcal/mol) with FOT (target protein). It had nine hydrogen bonds. It interacted with the active residues of FOT (ASP 467, GLN 468, ARG 80, PRO 93, SER 95 and ALA 466). Punicalagin and corilagin may be potential as an anti-obesity drug.

Dina et al. (2019) found that stigmast-5-en-3-ol bound strongly with leptin receptor at the lowest binding energy (−6.1 kcal/mol). Stigmast-5-en-3-ol derived from the seed extract of *Foeniculum vulgare*. The molecular docking analysis was carried out using Autodock Vina. The results demonstrated that stigmast-5-en-3-ol had a significant action towards the anti-obesity mechanism. Thus, it could be a good candidate for treating obesity disease.

12.7 Conclusion

In summary, a computational biology approach will prove promising to help in discovering and designing novel, effective and more potent structure-based anti-obesity drugs with a better understanding of drug and receptor interactions.

References

Adeneye AA, Adeyemi OO, Agbaje EO (2010) Anti-obesity and antihyperlipidaemic effect of *Hunteria umbellata* seed extract in experimental hyperlipidaemia. Journal of Ethnopharmacology 130(2): 307-314

Ahmad N, Abbasi BH, Fazal H (2016) Effect of different *in vitro* culture extracts of black pepper (*Piper nigrum* L.) on toxic metabolites-producing strains. Toxicology and Industrial Health 32 (3): 500-506

Akhila S, Anu V, Preeja GP (2020) An *in silico* approach to curtail the action of obesity proteins with phytoconstituents International Journal of Current Advanced Research 9(3): 21479-21481

Alarcon-Aguilar FJ, Zamilpa A, Perez-Garcia MD, Almanza-Perez JC, Romero-Nunez, E, Campos-Sepulveda EA, Roman-Ramos R (2007) Effect of *Hibiscus sabdariffa* on obesity in MSG mice. Journal of Ethnopharmacology 114(1): 66-71

Alam M, Khan H, Samiullah, L, Siddique KM (2012) A review on phytochemical and pharmacological studies of Kundur (*Boswellia serrata* Roxb ex Colebr.)-A Unani drug. Journal of Applied Pharmaceutical Science 2(3): 148-156

Ali BH, Blunden G, Tanira MO, Nemmar A (2008) Some phytochemical, pharmacological and toxicological properties of ginger (*Zingiber officinale* Roscoe): a review of recent research. Food and Chemical Toxicology 46(2): 409-420

Alias N, Ghazali SN, Hamid AA (2020) Molecular docking and dynamic simulation of astragalin reveals inhibitory potential against pancreatic lipase. International Journal of Allied Health Sciences 4(2): 1174-1189

Alwan H, Viswanathan B, Paccaud F, Bovet P (2011) Is accurate perception of body image associated with appropriate weight-control behavior among adolescents of the Seychelles. Journal of Obesity 2011: 1-9

Audu A, Egwim EC, Adeyemi HY (2014) Hypolipidemic properties of four varieties of eggplants (*Solanum melongena L.)* International Journal of Pharmaceutical Science Invention 3(8): 47-54

Austin A, Herbals R (2008) A review on Indian sarsaparilla, *Hemides musindicus* (L.)R. Br. Journal of Biological Sciences 8(1), 1-12

Azman KF, Amom Z, Azlan A, Esa NM, Ali RM, Shah ZM, Kadir KKA (2012) Anti-obesity effect of *Tamarindu sindica* L. pulp aqueous extract in high-fat diet-induced obese rats. Journal of Natural Medicines 66(2) 333-342

Bhat SK, Sarpangala M, Ashwin D (2017) Anti-lipidemic activity of arecanut, *Areca catechu* L.: A valuable herbal medicine. International Journal of Herbal Medicine 5(1): 35-38

Bordia A, Verma SK, Srivastava KC (1998) Effect of garlic (*Allium sativum*) on blood lipids, blood sugar, fibrinogen and fibrinolytic activity in patients with coronary artery disease. Prostaglandins, Leukotrienes and Essential Fatty Acids 58(4): 257-263

Bracesco N, Sanchez AG, Contreras V, Menini T, Gugliucci A (2011) Recent advances on *Ilex paraguariensis* research: mini review. Journal of Ethnopharmacology 136(3): 378-384

Brinkhaus B, Lindner M, Schuppan D, Hahn EG (2000) Chemical, pharmacological and clinical profile of the East Asian medical plant *Centella aslatica*. Phytomedicine 7(5): 427-448

Bunawan H, Bunawan SN, Baharum SN, Noor NM (2015) *Sauropus androgynus* (L.) Merr. induced bronchiolitis obliterans: from botanical studies to toxicology. Evidence-Based Complementary and Alternative Medicine 2015: 1-7

Cercato LM, White PA, Nampo FK, Santos MR, Camargo EA (2015) A systematic review of medicinal plants used for weight loss in Brazil: Is there potential for obesity treatment? Journal of Ethnopharmacology 176: 286-296

Chandrasekaran CV, Vijayalakshmi MA, Prakash K, Bansal VS, Meenakshi J, Amit A (2012) Herbal approach for obesity management. American Journal of Plant Sciences 3(7): 1003

Choi H, Eo H, Park K, Jin M, Park EJ, Kim SH, Kim S (2007) A water-soluble extract from *Cucurbita moschata* shows anti-obesity effects by controlling lipid metabolism in a high fat diet-induced obesity mouse model. Biochemical and Biophysical Research Communications 359(3): 419-425

Dey A (2011) *Achyranthesaspera L*: phytochemical and pharmacological aspects. International Journal of Pharmaceutical Sciences Review and Research 9(2): 72-82

Dickel ML, Rates SMK, Ritter MR (2007) Plants popularly used for loosing weight purposes in Porto Alegre, South Brazil. Journal of Ethnopharmacology 109(1): 60-71

Dina AG, Mazin YA, Reem M, Tasnim OE, Yosra AM, Tarig MH, Ali AS (2019) Investigation of anti-obesity activity of ethanolic extract of *Foeniculum vulgare* seeds, *in vivo and in silico* models. World Journal of Pharmacy and Pharmaceutical Sciences 8(11): 111-124

Governa P, Baini G, Borgonetti V, Cettolin G, Giachetti D, Magnano AR, Biagi M (2018) Phytotherapy in the management of diabetes: A Review. Molecules 23(1): 105

Falowo AB, Mukumbo FE, Idamokoro EM, Lorenzo JM, Afolayan A, Muchenje V (2018) Multi-functional application of *Moringa oleifera Lam.* in nutrition and animal food products: A review. Food Research International 106: 317-334

Gandhi SP, Lokhande KB, Swamy VK, Nanda RK, Chitlange SS (2019) Computational data of phytoconstituents from *Hibiscus rosa-sinensis* on various anti-obesity targets. Data in Brief 24: 103994

González-Molina E, Domínguez-Perles R, Moreno DA, García-Viguera C (2010) Natural bioactive compounds of *Citrus limon* for food and health. Journal of Pharmaceutical and Biomedical Analysis 51(2): 327-345

Heymsfield SB, Allison DB, Vasselli JR, Pietrobelli A, Greenfield D, Nunez C (1998) *Garcinia cambogia* (hydroxycitric acid) as a potential anti-obesity agent: a randomized controlled trial. JAMA 280(18): 1596-1600

Ikarashi N, Toda T, Okaniwa T, Ito K, Ochiai W, Sugiyama K (2011) Anti-obesity and anti-diabetic effects of acacia polyphenol in obese diabetic mice fed high-fat diet. Evidence-Based Complementary and Alternative Medicine 2011: 1-10

Jayaram B (2004) Active site prediction server. http://www.scfbio-iitd.res.in/dock/ActiveSite.jsp; accessed on 18th September 2020

Kamalakkannan S, Rajendran R, Venkatesh RV, Clayton P, Akbarsha MA (2010) Antiobesogenic and antiatherosclerotic properties of *Caralluma fimbriata* extract. Journal of Nutrition and Metabolism 2010: 1-6

Karmase A, Birari R, Bhutani KK (2013) Evaluation of anti-obesity effect of *Aegle marmelos* leaves. Phytomedicine 20(10): 805-812

Khambalia AZ, Seen LS (2010) Trends in overweight and obese adults in Malaysia (1996-2009): a systematic review. Obesity Reviews 11(6): 403-412

Kim H, Bartley GE, Arvik T, Lipson R, Nah SY, Seo K, Yokoyama W (2014) Dietary supplementation of chardonnay grape seed flour reduces plasma cholesterol concentration, hepatic steatosis, and abdominal fat content in high-fat diet-induced obese hamsters. Journal of Agricultural and Food Chemistry 62(8): 1919-1925

Kim MJ, Park M, Jeong MK, Yeo J, Cho WI, Chang PS, Lee J (2010) Radical scavenging activity and anti-obesity effects in 3T3-L1 preadipocyte differentiation of Ssuk (*Artemisia princeps Pamp.*) extract. Food Science and Biotechnology 19(2): 535-540

Klein G, Kim J, Himmeldirk K, Cao Y, Chen X (2007) Anti-diabetes and anti-obesity activity of *Lagerstroemia speciosa*. Evidence-Based Complementary and Alternative Medicine 4(4): 401-407

Krishna KL, Paridhavi M, Patel JA (2008) Review on nutritional, medicinal and pharmacological properties of Papaya (*Carica papaya Linn.*). Natural Product Radiance 7(4): 364-373

Kuppusamy UR, Das NP (1992) Effects of flavonoids on cyclic AMP phosphodiesterase and lipid mobilization in rat adipocytes. Biochemical Pharmacology 44: 1307-1315

Kuppusamy UR, Das NP (1994) Potentiation of beta adrenoceptor agonist-mediated lipolysis by quercetin and fisetin in isolated rat adipocytes. Biochemical Pharmacology 47: 521-519

Kwon CS, Sohn HY, Kim SH, Kim JH, Son KH, Lee JS, Kim JS (2003) Anti-obesity effect of *Dioscorea nipponica Makino* with lipase-inhibitory activity in rodents. Bioscience, Biotechnology, and Biochemistry 67(7): 1451-1456

Lee SH, Kim B, Oh MJ, Yoon J, Kim HY, Lee KJ., Choi KY (2011) *Persicaria hydropiper* (L.) spach and its flavonoid components, isoquercitrin and isorhamnetin, activate the Wnt/β-catenin pathway and inhibit adipocyte differentiation of 3T3-L1 cells. Phytotherapy Research 25(11): 1629-1635

Lee B, Kwon M, Choi JS, Jeong HO, Chung HY, Kim HR (2015) Kaempferol isolated from *Nelumbo nucifera* inhibits lipid accumulation and increases fatty acid oxidation signaling in adipocytes. Journal of Medicinal Food 18(12): 1363-1370

Macarron R (2006) Critical review of the role of HTS in drug discovery. Drug Discovery Today. 11: 277-279

Manchanda S, Kaur G (2017) *Withania somnifera* leaf alleviates cognitive dysfunction by enhancing hippocampal plasticity in high fat diet induced obesity model. BMC Complementary and Alternative Medicine 17(1): 136

Mohammed A, Al-Numair KS, Balakrishnan A (2015) Docking studies on the interaction of flavonoids with fat mass and obesity associated protein. Pakistan Journal of Pharmaceutical Sciences 28(5): 1647-1653

Moro CO, Basile G (2000) Obesity and medicinal plants. Fitoterapia 71: S73-S82

Munhoz A, Frode TS (2018) Isolated compounds from natural products with potential antidiabetic activity-a systematic review. Current Diabetes Reviews 14(1): 36-106

Nakagawa K, Kishida H, Arai N, Nishiyama T, Mae T (2004) Licorice flavonoids suppress abdominal fat accumulation and increase in blood glucose level in obese diabetic KK-Ay mice. Biological and Pharmaceutical Bulletin 27(11): 1775-1778

National Health and Morbidity Survey (NHMS) (2015) https://www.moh.gov.my/moh/resources/nhmsreport2015vol2.pdf; accessed on 18th September 2019

Nerurkar PV, Lee YK, Nerurkar VR (2010) *Momordica charantia* (bitter melon) inhibits primary human adipocyte differentiation by modulating adipogenic genes. BMC Complementary and Alternative Medicine 10(1): 34

Nishiyama T, Mae T, Kishida H, Tsukagawa M, Mimaki Y, Kuroda M, Kitahara M (2005) Curcuminoids and sesquiterpenoids in turmeric (*Curcuma longa L.*) suppress an increase in blood glucose level in type 2 diabetic KK-Ay mice. Journal of Agricultural and Food Chemistry 53(4): 959-963

Pothuraju R, Sharma RK, Chagalamarri J, Jangra S, Kumar Kavadi P (2014) A systematic review of *Gymnema sylvestre* in obesity and diabetes management. Journal of the Science of Food and Agriculture 94(5): 834-840

Raju J, Bird RP (2006) Alleviation of hepatic steatosis accompanied by modulation of plasma and liver TNF-α levels by *Trigonella foenumgraecum* (fenugreek) seeds in Zucker obese (fa/fa) rats. International Journal of Obesity 30(8): 1298

Rédei D, Kúsz N, Jedlinszki N, Blazsó G, Zupkó I, Hohmann, J (2018) Bioactivity-guided investigation of the anti-inflammatory activity of *Hippophae Rhamnoides* fruits. Planta Medica 84(1): 26-33

Sanner MF (1999) Python: A programming language for software integration and development. Journal of Molecular Graphics and Modelling 17: 57-61

Seto SW, Lam TY, Tam HL, Au ALS, Chan SW, Wu JH, Leung PS (2009) Novel hypoglycemic effects of *Ganoderma lucidum* water-extract in obese/diabetic (+ db/+ db) mice. Phytomedicine 16(5): 426-436

Sharma V, Singh M (2012) Alterations induced by N-Nitrosodimethylamine and ethanolic root extract of *Operculina turpethum* in serum lipid profile of male albino mice. Asian Journal of Pharmaceutical Clinical Research 5(3): 69-73

Singh R, Sharma, PK, Malviya R (2011) Pharmacological properties and ayurvedic value of Indian buch plant (*Acorus calamus*): a short review. Advances in Biological Research 5(3): 145-154

Sivashanmugam T, Muthukrishnan S, Umamaheswari M, Asokkumar K, Subhadradevi V, Jagannath P, Madeswaran A (2013) Discovery of potential cholesterol esterase inhibitors using *in silico* docking studies. Bangladesh Journal of Pharmacology 8(3): 223-229

Sudeep H, Prasad KS (2014) Computational studies on the anti-obesity effect of polyphenols from pomegranate leaf. Journal of Chemical and Pharmaceutical Research 6(9): 278-281

Sung YY, Kim DS, Kim HK (2015) *Akebia quinata* extract exerts anti-obesity and hypolipidemic effects in high-fat diet-fed mice and 3T3-L1 adipocytes. Journal of Ethnopharmacology 168: 17-24

Tanaka M, Misawa E, Ito Y, Habara N, Nomaguchi K, Yamada M, Higuchi R (2006) Identification of five phytosterols from *Aloe vera* gel as anti-diabetic compounds. Biological and Pharmaceutical Bulletin 29(7): 1418-1422

Vasudeva N, Yadav N, Sharma SK (2012) Natural products: a safest approach for obesity. Chinese Journal of Integrative Medicine 18(6): 473-480

World Health Organization (2015) https://www.who.int/news-room/fact-sheets/detail/obesity-and-overweight; accessed on 18th September 2019

Yoshikawa M, Shimoda H, Nishida N, Takada M, Matsuda H (2002) *Salacia reticulata* and its polyphenolic constituents with lipase inhibitory and lipolytic activities have mild antiobesity effects in rats. The Journal of Nutrition 132(7): 1819-1824

Yuliana ND, Iqbal M, Jahangir M, Wijaya CH, Korthout H, Kottenhage M, Verpoorte R (2011) Screening of selected Asian spices for anti-obesity-related bioactivities. Food Chemistry 126(4): 1724-1729

Zoy N, Ahmed D, Rehman HM, Qamar MT, Froeyen M, Ahmad S, Mirza MU (2019) In vitro antidiabetic, anti-obesity and antioxidant analysis of Ocimum Basilicum aerial biomass and in silico molecular docking simulations with alpha-amylase and lipase enzymes. Biology 8(4): 92

The manufacturer's authorised representative in the EU is Springer Nature Customer Service Centre GmbH, Europaplatz 3, 69115 Heidelberg, Germany. If you have any concerns regarding our products, please contact ProductSafety@springernature.com

Printed and bound by CPI Group (UK) Ltd, Croydon, CR0 4YY

30/09/2026

02225266-0002